PRACTICAL DESIGN AND APPLICATIONS OF MEDICAL DEVICES

PRACTICAL DESIGN AND APPLICATIONS OF MEDICAL DEVICES

Edited by

DILBER UZUN OZSAHIN

Department of Medical Diagnostic Imaging, College of Health Science, University of Sharjah, Sharjah, United Arab Emirates; Research Institute for Medical and Health Sciences, University of Sharjah, Sharjah, United Arab Emirates; Operational Research Center in Healthcare, Near East University, Nicosia/TRNC, Mersin 10, Turkey

ILKER OZSAHIN

Department of Radiology, Brain Health Imaging Institute, Weill Cornell Medicine, New York, NY, United States; Operational Research Center in Healthcare, Near East University, Nicosia/TRNC, Mersin 10, Turkey

ACADEMIC PRESS

An imprint of Elsevier

Notices
Knowledge and best practice in this field are constantly changing. As new research and experience broaden our understanding, changes in research methods, professional practices, or medical treatment may become necessary.

Practitioners and researchers must always rely on their own experience and knowledge in evaluating and using any information, methods, compounds, or experiments described herein. In using such information or methods they should be mindful of their own safety and the safety of others, including parties for whom they have a professional responsibility.

To the fullest extent of the law, neither the Publisher nor the authors, contributors, or editors, assume any liability for any injury and/or damage to persons or property as a matter of products liability, negligence or otherwise, or from any use or operation of any methods, products, instructions, or ideas contained in the material herein.

ISBN: 978-0-443-14133-1

For Information on all Academic Press publications
visit our website at https://www.elsevier.com/books-and-journals

Publisher: Mara Conner
Acquisitions Editor: Carrie Bolger
Editorial Project Manager: Fernanda Oliveira
Production Project Manager: Anitha Sivaraj
Cover Designer: Mark Rogers

Typeset by MPS Limited, Chennai, India

Working together
to grow libraries in
developing countries

www.elsevier.com • www.bookaid.org

Contents

List of contributors

Obada Abid
Department of Biomedical Engineering, Near East University, Nicosia/TRNC, Mersin 10, Turkey

Belal J.N. Abuamsha
Department of Biomedical Engineering, Near East University, Nicosia/TRNC, Mersin 10, Turkey

Alaa M.Y. Abuedia
Department of Biomedical Engineering, Near East University, Nicosia/TRNC, Mersin 10, Turkey

Fadel Alayouti
Department of Biomedical Engineering, Near East University, Nicosia/TRNC, Mersin 10, Turkey

Abdullah Alchoib
Department of Biomedical Engineering, Near East University, Nicosia/TRNC, Mersin 10, Turkey

Mohammad Aldakhil
Department of Biomedical Engineering, Near East University, Nicosia/TRNC, Mersin 10, Turkey

Ayman Aleter
Department of Biomedical Engineering, Near East University, Nicosia/TRNC, Mersin 10, Turkey

Majd Issam Ali
Department of Biomedical Engineering, Near East University, Nicosia/TRNC, Mersin 10, Turkey

Ghazy Aljammal
Department of Biomedical Engineering, Near East University, Nicosia/TRNC, Mersin 10, Turkey

Riad Alsabbagh
Department of Biomedical Engineering, Near East University, Nicosia/TRNC, Mersin 10, Turkey

Moayad Alshobaki
Department of Biomedical Engineering, Near East University, Nicosia/TRNC, Mersin 10, Turkey

Abdulraheem Alsiba
Department of Biomedical Engineering, Near East University, Nicosia/TRNC, Mersin 10, Turkey

Samer M.Y. Altartoor
Department of Biomedical Engineering, Near East University, Nicosia/TRNC, Mersin 10, Turkey

Saleem Attili
Department of Biomedical Engineering, Near East University, Nicosia/TRNC, Mersin 10, Turkey

Bahaaeddin A.T. Bader
Department of Biomedical Engineering, Near East University, Nicosia/TRNC, Mersin 10, Turkey

Ali Mohsen Banat
Department of Biomedical Engineering, Near East University, Nicosia/TRNC, Mersin 10, Turkey

Janee Dinesh Barot
Department of Biomedical Engineering, Near East University, Nicosia/TRNC, Mersin 10, Turkey

Dawda Cham
Department of Biomedical Engineering, Near East University, Nicosia/TRNC, Mersin 10, Turkey

Basil Bartholomew Duwa
Operational Research Center in Healthcare, Near East University, Nicosia/TRNC, Mersin 10, Turkey; Department of Biomedical Engineering, Near East University, Nicosia/TRNC, Mersin 10, Turkey

Mohamad Sharaf Eddin
Department of Biomedical Engineering, Near East University, Nicosia/TRNC, Mersin 10, Turkey

David Edward
Department of Biochemistry, Faculty of Basic Medical Sciences, University of Jos, Jos, Nigeria

Nosaiba Elhassan Eldasougi
Department of Biomedical Engineering, Near East University, Nicosia/TRNC, Mersin 10, Turkey

Khaldun Elsafdy
Department of Biomedical Engineering, Near East University, Nicosia/TRNC, Mersin 10, Turkey

Declan Ikechukwu Emegano
Operational Research Center in Healthcare, Near East University, Nicosia/TRNC, Mersin 10, Turkey; Department of Biomedical Engineering, Near East University, Nicosia/TRNC, Mersin 10, Turkey

James Gambu
Department of Medical Microbiology and Clinical Microbiology, Near East University, Nicosia/TRNC, Mersin 10, Turkey

Cemalettin Yağiz Günaşti
Department of Biomedical Engineering, Near East University, Nicosia/TRNC, Mersin 10, Turkey

Omar Haider
Department of Biomedical Engineering, Near East University, Nicosia/TRNC, Mersin 10, Turkey

Mohamad Khir Hamdan
Department of Biomedical Engineering, Near East University, Nicosia/TRNC, Mersin 10, Turkey

Noor Hamzah
Department of Biomedical Engineering, Near East University, Nicosia/TRNC, Mersin 10, Turkey

Ibrahim Hashi
Department of Biomedical Engineering, Near East University, Nicosia/TRNC, Mersin 10, Turkey

Abdulsamad Hassan
Department of Biomedical Engineering, Near East University, Nicosia/TRNC, Mersin 10, Turkey

Shahida Hassan
Department of Biomedical Engineering, Near East University, Nicosia/TRNC, Mersin 10, Turkey

Abdelrahman Himaid
Department of Biomedical Engineering, Near East University, Nicosia/TRNC, Mersin 10, Turkey

Osama Haj Hussein
Department of Biomedical Engineering, Near East University, Nicosia/TRNC, Mersin 10, Turkey

Ismail Ibrahim
Department of Biomedical Engineering, Near East University, Nicosia/TRNC, Mersin 10, Turkey

Bartholomew Idoko
Department of Physics, Advance Educational Consult and Research Institute, Abuja, Nigeria

John Bush Idoko
Applied Artificial Intelligence Research Center, Department of Computer Engineering, Near East University, Nicosia/TRNC, Mersin 10, Turkey

Deborah Ishimwe
Department of Biomedical Engineering, Near East University, Nicosia/TRNC, Mersin 10, Turkey

Mohamed Ismail
Department of Biomedical Engineering, Near East University, Nicosia/TRNC, Mersin 10, Turkey

Lama Khorzom
Department of Biomedical Engineering, Near East University, Nicosia/TRNC, Mersin 10, Turkey

Mubarak Taiwo Mustapha
Operational Research Center in Healthcare, Near East University, Nicosia/TRNC,
Mersin 10, Turkey; Department of Biomedical Engineering, Near East University,
Nicosia/TRNC, Mersin 10, Turkey

Mohamad Naesa
Department of Biomedical Engineering, Near East University, Nicosia/TRNC, Mersin 10,
Turkey

Mohammed Al Obied
Department of Biomedical Engineering, Near East University, Nicosia/TRNC, Mersin 10,
Turkey

Efe Precious Onakpojeruo
Operational Research Center in Healthcare, Near East University, Nicosia/TRNC,
Mersin 10, Turkey; Department of Biomedical Engineering, Near East University,
Nicosia/TRNC, Mersin 10, Turkey

Ilker Ozsahin
Operational Research Center in Healthcare, Near East University, Nicosia/TRNC,
Mersin 10, Turkey; Department of Radiology, Brain Health Imaging Institute, Weill
Cornell Medicine, New York, NY, United States

Angelique Rwiyereka
Department of Biomedical Engineering, Near East University, Nicosia/TRNC, Mersin 10,
Turkey

Emad Shoaireb
Department of Biomedical Engineering, Near East University, Nicosia/TRNC, Mersin 10,
Turkey

Mohammed Skaik
Department of Biomedical Engineering, Near East University, Nicosia/TRNC, Mersin 10,
Turkey

Waqqas Atiq Ur Raheman Subedar
Department of Biomedical Engineering, Near East University, Nicosia/TRNC, Mersin 10,
Turkey

Galaya Tirah
Department of Medical Biology and Genetics, Near East University, Nicosia/TRNC,
Mersin 10, Turkey

Natacha Usanase
Operational Research Center in Healthcare, Near East University, Nicosia/TRNC,
Mersin 10, Turkey; Department of Biomedical Engineering, Near East University,
Nicosia/TRNC, Mersin 10, Turkey

Dilber Uzun Ozsahin
Department of Medical Diagnostic Imaging, College of Health Science, University of
Sharjah, Sharjah, United Arab Emirates; Research Institute for Medical and Health
Sciences, University of Sharjah, Sharjah, United Arab Emirates; Operational Research
Center in Healthcare, Near East University, Nicosia/TRNC, Mersin 10, Turkey

Tevfik Yavuz
Department of Biomedical Engineering, Near East University, Nicosia/TRNC, Mersin 10, Turkey

Mohammad Eyad Osama Yousef
Department of Biomedical Engineering, Near East University, Nicosia/TRNC, Mersin 10, Turkey

About the editors

Dr. Dilber Uzun Ozsahin is an associate professor of College of Health Science, Medical Diagnostic Imaging Department, at the University of Sharjah since 2021. She graduated from the Department of Physics at Çukurova University in 2006. She worked at CERN, Geneva, during 2008−10 for her master's thesis. She completed her PhD studies on medical imaging for breast cancer at the Universitat Autònoma de Barcelona, Spain, in 2014. Between 2015 and 2016, she worked at the Massachusetts General Hospital and Harvard Medical School as a postdoctoral researcher. She has been working as an associate professor in the Biomedical Engineering Department at Near East University and the DESAM Institute between 2016 and 2021. She and her group are the first group to apply operational research tools to solve biomedical and medical problems. She is the director of the Operational Research Center in Healthcare, Near East University, since 2020. Also, she is the board member of the DESAM Institute, Near East University, since 2018. Currently, she is establishing a new research group named "Operational Research and Integrated Artificial Intelligence in Healthcare" at the University of Sharjah.

Associate Professor Ilker Ozsahin is a physicist and received his bachelor's degree in physics and PhD in medical imaging from Çukurova University. He has experience in medical imaging devices such as PET, SPECT, and Compton Camera, including modeling, simulation, and characterization. After earning his PhD, he worked at Harvard Medical School and Massachusetts General Hospital for 2 years as a postdoctoral fellow. Also, he worked as a visiting postdoctoral fellow at the University of Macau on multipinhole SPECT collimator design and implementation for brain imaging, as well as cardiac and small animal imaging by using adaptive collimators. Recently, he worked at CERN for HCAL upgrade studies in the CMS experiment. He has been also working on AI applications and operational research in healthcare. Currently, he is a visiting fellow at Cornell University and a faculty in the Biomedical Engineering Department at Near East University.

Preface

"Practical Design and Applications of Medical Devices" is a comprehensive guide that explores the captivating realm of medical device design and its practical implications in healthcare. It is specifically crafted to offer engineers, researchers, healthcare professionals, and students valuable insights into the process of designing, developing, and implementing medical devices that directly impact patient care and well-being.

In today's fast-paced technological landscape, medical devices play a vital role in the diagnosis, monitoring, and treatment of various medical conditions. Ranging from simple diagnostic tools to intricate implantable devices, medical device design encompasses multiple disciplines such as engineering, biomedical sciences, materials science, and regulatory affairs. The objective of this book is to bridge the gap between theory and practice by presenting real-world examples, case studies, and practical guidelines to help readers navigate the complexities of designing effective and safe medical devices.

The chapters follow a step-by-step approach, covering key considerations at each stage of the design and development process. The book begins by exploring the regulatory landscape, providing insights into the requirements and standards that govern the design, manufacturing, and marketing of medical devices. Understanding the regulatory framework is essential to ensure compliance and prioritize patient safety.

Subsequently, the book delves into the design process itself, emphasizing the significance of user-centered design principles, human factors engineering, and usability testing. By placing end-users at the heart of the design process, the aim is to create medical devices that are intuitive, ergonomic, and tailored to meet the specific needs of patients and healthcare professionals.

Throughout the book, critical design challenges and considerations for various types of medical devices are addressed, including diagnostic devices, therapeutic devices, monitoring devices, and assistive devices. Topics covered include materials selection, sterilization techniques, risk management, software development, and validation processes. The book also explores emerging trends and technologies such as wearable devices, telemedicine, and the integration of artificial intelligence in medical devices, which have the potential to revolutionize healthcare delivery.

Furthermore, the book examines essential aspects such as clinical trials, postmarket surveillance, and regulatory compliance. These chapters offer valuable insights into the steps required to ensure the safety, efficacy, and market success of medical devices.

The book has been compiled through collaboration with experts from academia, industry, and healthcare. Their contributions, combined with our own experiences in the field, have shaped the content of this comprehensive resource, offering diverse perspectives on the practical applications of medical device design.

It is our sincere hope that this book serves as a valuable reference for professionals and students interested in the dynamic and ever-evolving field of medical device design. Our ultimate goal is to empower readers to develop innovative, safe, and effective medical devices that contribute to improving patient care and enhancing overall quality of life.

Dilber Uzun Ozsahin

Introduction

Dilber Uzun Ozsahin[1,2], Basil Bartholomew Duwa[2,3] and Ilker Ozsahin[2,4]

[1]Department of Medical Diagnostic Imaging, College of Health Science, University of Sharjah, Sharjah, United Arab Emirates
[2]Operational Research Center in Healthcare, Near East University, Nicosia/TRNC, Mersin 10, Turkey
[3]Department of Biomedical Engineering, Near East University, Nicosia/TRNC, Mersin 10, Turkey
[4]Brain Health Imaging Institute, Department of Radiology, Weill Cornell Medicine, New York, NY, United States

Recent years have witnessed explosive expansion in the healthcare industry, resulting in significant increases in both revenue and employment. Up until now, the only way to determine whether or not someone had an illness or an abnormality in their body was to take them to the hospital for an examination. The majority of patients were required to remain hospitalized for the entirety of their course of treatment. This not only led to higher healthcare costs but also put pressure on the healthcare facilities available in more rural and distant areas. Because of the technical progress that has been made over the past few years, it is now possible to diagnose a wide range of ailments and monitor one's health with the help of small gadgets. These devices are the result of the increased need for simple medical equipment.

Furthermore, not only have medical gadgets-based Internet of Things (IoT) made independence more possible, but they have also expanded people's capacities for engaging with the world around them in new ways. The IoT became an important factor in the advancement of global communication thanks to the development of cutting-edge protocols and algorithms. It allows a huge variety of gadgets, wireless sensors, electrical equipment, and home appliances to be connected to the internet. Similarly, over the years, a significant amount of research has been conducted in the field of healthcare services and the technical advancement of these services. To be more specific, the IoT has demonstrated the potential applicability of connecting a variety of medical devices, sensors, and professionals in the medical field to provide excellent medical services in remote places.

This book presents the practical application and development of advanced medical devices applied in disease diagnosis, treatment, and prevention. This is being done to promote operational efficiency in the healthcare industry, as well as improve patient safety, lower overall healthcare costs, broaden access to healthcare services, and improve accessibility.

CHAPTER ONE

Design of interactive neural input device for arm prosthesis

Dilber Uzun Ozsahin[1,2,3], Basil Bartholomew Duwa[3,4], John Bush Idoko[5], Galaya Tirah[6], Abdullah Alchoib[4], Alaa M.Y. Abuedia[4], Moayad Alshobaki[4], Deborah Ishimwe[4] and Ilker Ozsahin[3,7]

[1]Department of Medical Diagnostic Imaging, College of Health Science, University of Sharjah, Sharjah, United Arab Emirates
[2]Research Institute for Medical and Health Sciences, University of Sharjah, Sharjah, United Arab Emirates
[3]Operational Research Center in Healthcare, Near East University, Nicosia/TRNC, Mersin 10, Turkey
[4]Department of Biomedical Engineering, Near East University, Nicosia/TRNC, Mersin 10, Turkey
[5]Applied Artificial Intelligence Research Center, Department of Computer Engineering, Near East University, Nicosia/TRNC, Mersin 10, Turkey
[6]Department of Medical Biology and Genetics, Near East University, Nicosia/TRNC, Mersin 10, Turkey
[7]Department of Radiology, Brain Health Imaging Institute, Weill Cornell Medicine, New York, NY, United States

Contents

Practical Design and Applications of Medical Devices.
DOI: https://doi.org/10.1016/B978-0-443-14133-1.00006-9

1

1.1 Introduction

Upper extremities form a large part of the injuries treated in emergency departments across the globe. Most of them happen at home, at work, or while playing sports. Given that, almost all of our daily tasks depend on hand manipulation, serious hand injuries can truly be devastating. The effects of such events can lead to long-term injuries, which often impact the mental and social state, resulting in difficult social reintegration [1].

By the late 1960s, many joint and grip styles were capable of driving and controlling pneumatic prostheses. The regulation, however, was unreliable and not strong enough, requiring the patient's unique anatomical characteristics, dexterity, and cognitive effort. With state-based control, myoelectric control systems have attempted to overcome these challenges. As a result, unlike the single degree of freedom (DOF), the patient can operate the prosthesis utilizing two control sites. A cocontraction of the muscles under the two recording sites changed the control state of the prosthesis when it was necessary to regulate a certain joint or grip form. On the market for dexterous prosthetics, this very cognitively demanding device is still dominant, mainly due to its robustness. Hand transplantation provides versatility, superior aesthetic appeal, and an integrated sensory feature as an alternative to prosthetic devices. However, lifelong immunosuppressant treatment, extended recovery, loss of grip power, and a high risk of complications are associated with this, leading to their potential rejection. Such concerns are further associated with very high expenses [1−3].

This analysis aims to present the current state of the art of practical, myoelectrically regulated upper limb prosthetic solutions, considering the latest rates of innovation in the field of prosthetics, a substantial rise in funding, and the number of new competitors on the market. The aim is to provide a literature review from both the medical and technological viewpoints of recent developments in representative hardware, control algorithms, and interfaces. Solutions and studies in the field of sensory feedback are outside the scope of this study but are highly relevant for prosthetic applications [2].

1.1.1 Passive functional hand prostheses

There are no moving parts in the passive prosthetic hand, so it can be used for holding, pushing, and pulling. Passive prostheses usually look like fingers, hands, and arms. They are lightweight, and while they do not have active mobility, they can assist a person's function by providing a surface to carry or grasp objects. Passive suits can be stuffed with specially coated high-precision silicone that nearly simulates a voice arm, hand, and fingers, or a reduced manufacturing glove. Multiposition joints are frequently used in conjunction with a passive prosthesis to increase human function by allowing the user to adjust the shoulder, elbow, wrist, or finger joints. For example, the sound hand may be placed at a certain angle with a multiposition shoulder joint, elbow, or wrist, making it easier to hold or carry the object. To allow high-definition restoration to catch small artifacts, multipositional finger joints can be moved into place. Patients have to decide what form of prosthesis they would like to use after an amputation that would be ideal for their activity level and lifestyle [4–6]. It must be determined whether the arm prosthesis will be more practical or whether it will emphasize a natural appearance. For example, if greater emphasis is put on the appearance and comfort of the wearer, there is a range of visually attractive passive arm prosthetics to offer. However, when managing objects, their functionality is restricted to that of a basic counter-support. Several features to mimic the natural hand in great detail when designing Otto Bock cosmetic hands, consisting of an inner hand and an easy-to-care-for cosmetic glove, were included. For custom adaptation, 43 models for children, women, and men are available. Each model is available in 18 natural shades to choose from. Your prosthetist would be able to give you more details about the use of hands with the passive method.

1.1.2 Body-powered prosthesis

Either a functional prosthesis or electricity is powered by the body. Body-powered devices are operated using cable and brace systems to allow the patient to pull the cable to open or close the terminal device (hand, hook, or prehensor), much the way a bicycle handbrake system operates using body movements (shifting the shoulders or arm). Voluntary opening or voluntary-closing is mechanical body-operated terminal equipment. Voluntary opening means that by applying force through their cable system, the users must open the terminal unit. With the help of rubber bands, the terminal mechanism then closes on its own, limiting the grip strength of the device to the strength of the rubber bands. Force must be exerted to close it instead of opening it with a voluntary closing terminal system, rendering the grip strength not on the strength of the rubber bands, but on the strength of the person using them [4].

1.1.3 Amputation

The word amputation refers to the process in which a part of the body is separated, and the term prosthesis refers to the artificial device that is used as a substitute for a lost or amputated part of the body. The part that remains intact after the amputation is called the rhizome, while the part placed over the limb is called the shirt name [2,3]. A person who has had an amputation can return to his normal life if he gets the appropriate care, and then he can practice many of the actions that a normal person does because of the capabilities provided by the prosthetic and orthotic devices. This is demonstrated in Fig. 1.1.

After the amputation, the patient enters into a complex and unique psychological state due to the absence of the limb or part of it since some nerves are cut during the amputation process, causing the patient to feel that the limb is still present. It is possible to feel the limb is in a difficult position and pain in the rhizome. With time this imaginary sense of limb or pain decreases.

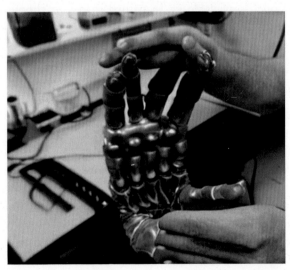

Figure 1.1 Prosthetic arm in motion. *Image is credited to Cleveland Clinic Center for Medical Art & Photography.*

There are various causes of amputation of the upper limb, including pathological and others. The following are the leading causes of amputation based on the statistics in America between 1988 and 1996 [5]:

1. Diseases resulting from blood vessels (dysvascular): This is the most common reason in developed countries because people live longer as a result of the high level of good living and healthcare. Vascular diseases begin to appear in the distal extremities from the blood pumping center, as they suffer from a decrease in the quantities of blood reaching them. The amputation of the limbs becomes a sure way of getting rid of the dead tissue. The percentage of cases for this type of amputation are 3% for the upper limb and 97% for the lower limb.

2. Accident injuries (trauma): Accidents are important causes of amputation. In developed countries, as a result of progress in existing technology, accidents take place in factories, and traffic accidents happen, in general. As for the third world countries, the first cause of accidents that lead to amputation is wars. In this type of amputation, the process should be delayed as much as possible because it is difficult to determine the viability of the affected tissues, especially when they are in distant extremities such as the feet. The percentage of cases of this type of amputation are 68.6% for the upper limb and 31% for the lower limb.

3. Cancer: If radiotherapy does not succeed, amputation becomes a sure solution. In this case the amputation must be through or above the joint adjacent to the swelling side closest to the human torso. The percentage of cases of this type of amputation are 23.9% for the upper limb and 76.1% for the lower limb.

4. Congenital cases (congenital): Studies show that out of every 100,000 births, there are 26 cases in which a limb is completely lost congenitally. The percentage of cases of this type of amputation are 58.5% for the upper limb and 41.5% for the lower limb [6].

1.2 Methodology

Prosthetic hands have many types, including fixed and movable ones. Fixed ones are used to compensate for limb loss in a cosmetic way only and without presenting any movement. As for the movable hands, they can be classified according to the type of control, including mechanical and electrical, or electronic.

1.2.1 Electromyogram used in prosthetic arm control

The device specifications for the implantable myoelectric sensor (IMES) system are driven by the evaluation of both standard and innovative control strategies. The key factor preventing the production of more advanced prosthetic arms is not the arm processes themselves, but instead the challenge of locating adequate sources of power to control the multiple DOFs needed to replace a physical arm [7–11]. The design of an IMES process that converts a subcutaneous (no cables) magnetic connection makes it possible to produce different control sources through the recording at their source prosthetic sensors with low interelectrode intermodulation values and hence a high sense of autonomy between sources. The most widely employed biosignal in the regulation of external device prosthetic materials is the electromyogram of artificial limbs [7]. As a logical outcome of muscle tissue stimulation, the electromyograph is produced with a range of electrolytic capacitor processes that can be easily identified and intensified. To decode user behavior to decide which sensors to move throughout the implants, intensified electromyogram signals can be transmitted to the implant processor for more analysis, i.e., how these inputs (signified electromyograms) connect to the outputs (prosthesis motors) is a multiple-input-multiple-output issue. In recent estimates, the World Health Organization estimated that nearly 15% of the world's population is disabled, with half of them unable to obtain healthcare. The total number of amputees and patients with limb dysfunction is rising due to different political, fiscal, scientific, and demographic factors. There are more than 10 million amputees globally, 30% of which are arm amputees [7]. While prosthetic limbs have been around for decades, in terms of function and contact with the environment, they are not very normal [5]. They require an invasive surgical procedure to be performed. The basic purpose of such sophisticated treatments, according to the John Hopkins Applied Physics Laboratory, is to reassign nerves and enable amputees to operate their prosthetic devices by simply thinking about the activity they need to accomplish.

The most common method is to employ an EEG (electroencephalography) unit, which captures subjects' cerebrum waves while they are thinking about something or looking at something. These readings are consequently changed over into mandates for the arm. The psyche is constrained by electrical waves recorded in the cerebrum that produce electrochemical driving forces of different frequencies that can be caught by an electroencephalogram, as indicated by the author. First off, beta waves are created when an individual is concerned or apprehensive, with frequencies going from 13 to 60 Hz. At the point when an individual feels mentally calm, alpha waves are delivered at frequencies going from 7 to 13 Hz. Delta waves are then delivered when a living being is in a state of obviousness. Progressions in innovation have made it conceivable to handle these EEG frequencies and information straightforwardly, continuously utilizing a cerebrum-computer interface that is a mix of software and hardware [10].

1.2.2 System architecture

As displayed in Fig. 1.2, the proposed framework is coordinated into four major units: the processing unit (pattern identification), the EEG sensor input unit, the interface unit (smart network of sensors), and the electro-mechanical gadget (the arm).

The first device contains an array of powerful EEG sensors that capture brain waves and is coupled to a signal processing unit through low-power and secure Bluetooth technology. This gadget has an internal sampling rate of 2048 Hz, and 14 sensors are placed in the global 10–20 framework to cover the most relevant brain regions. This guarantees that all components of brain activity are thoroughly and effectively covered.

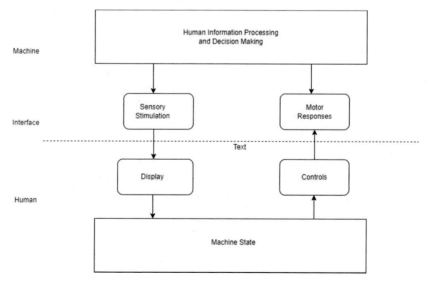

Figure 1.2 Flow design of the proposed device.

An intelligent sensor network, including temperature, skin pressure, ultrasonic vicinity sensors, accelerometers, potentiometers, strain measures, and gyroscopes, makes up the unit. The gadget's fundamental highlights empower the arm to respond to and communicate with the outside world. Furthermore, bidirectional coordination is required to send commands to the arm and receive feedback from the patient. Networking techniques and custom communication protocols are utilized in this actuator network and integrated sensor to permit the patient and arm to communicate effectively and monitor hand-over. Originally, the brain of the patient controls the arm; however, it could be moved to the arm to shield itself from injury [12].

The exhibited mind-controlled smart prosthetic arm can improve the quality of life for a large number of patients and their families due to its exceptional features. Its low-cost design would make it accessible to a wide range of people, including those who have limited or no access to specialized medical care.

1.2.2.1 Technical specifications

EEG neurofeedback is used to control the smart prosthetic arm, which is based on electromechanical technology. In terms of hardware, this design is categorized into four sections. In the EEG-based headgear, 14 EEG sensors are integrated via the data sampling unit. This device transforms EEG impulses into digital signals and sends them over Bluetooth to the processing unit.

Also inn the wireless networking device, a low-power Bluetooth module is connected to a Raspberry III microcontroller. The next section explains the communication protocol between the consumer (head-mounted EEG sensors) and the prosthetic arm's embedded CPU [13].

1.2.2.2 Computing and processing unit

The sampled EEG signals are sent to the computing and processing unit via the wireless communication unit, which is integrated within the arm and comprises a Raspberry Pi III microprocessor interfaced with an Arduino Mega microcontroller that controls the mechanical servo units implanted in the arm. The capacity to handle digitized EEG inputs is the processing unit's key feature. Its purpose is to link the headset's readings to a set of premeasured patterns linked to different mental states [14,15], which are gathered and encoded to indicate a series of tasks that the arm should do. This device also contains a lot of hand reflexes that are controlled by the arm's smart sensor network. Through fluid motions and rapid reflexes, it gives the arm a human-like behavior.

Finally, in the 3D model, the electro-mechanical machine has eight servos fitted. The 3D hand model is made up of a combination of high-strength lightweight materials that can withstand heavy impacts while still being delicate. It is implanted into the arm and connects servos to joints, allowing it to perform a range of actions. The servos are strategically arranged to avoid hardware and encourage intricate movements. In this configuration the mechanical device and the processing unit are connected via a microcontroller.

The power analysis of the proposed smart prosthetic arm is divided into two main sections:

a. When fully charged, the lithium battery in the EEG headset provides up to 12 hours of continuous use. It may take up to 4 hours to recharge a fully exhausted battery (after 12 hours of usage). However, charging time varies between 30 minutes and 2 hours when not fully charged. Before using it for more than 12 hours straight, it is also a good idea to refill it. The headset should not be used while it is being loaded for safety reasons.

b. The electromechanical system is made up of eight servo motors with rated powers of 0.5 and 3.5 W mounted on different parts of the pros-thetic arm (wrist, elbow, shoulder, and fingers). To provide three degrees of freedom, three servo motors with a rated strength of 3.5 W are positioned on the hand, elbow, and shoulder fields. The five fingers are also controlled by five servo motors with a rated power of 0.5 W apiece. As a result, the total power consumed by all servo engines is 13 W. To power all of the servo motors included in the arms, a 5 V battery with a 2.6 A output current is required. The wireless communication device consumes 0.165 W of power to operate when 50 mA is exchanged at 3.3 V. Accelerometers, potentiometers, strain gauges, and gyroscopes, as well as temperature, skin pressure, and proximity ultrasonic sensors, are widely used. A total current of 100 mA is absorbed by these sensors. The low-power single-board machine (computing and processing unit) consumes 0.1 W of electricity, or 30 mA. The machine requires a power source with a minimum output current of 2.8 A and a maximum output voltage of 5 V to power all of the above-mentioned units. Two lithium-ion batteries with a 10,000-mA capacity and a 2 A output current have been chosen. A charging circuit (through a USB connection), as well as a boost converter that delivers 5 V DC, are included. It is not recommended to use the arm while it is being charged because the battery has an energy loss of 80% on both ends. The arm will be active for a total of 7 hours with these two batteries, considering that the average daily movement of hands per person is 1 to 3 hours of continuous movement, depending on the daily work. The entire contraption may run for two days on end [15].

In the proposed method, it is necessary to distinguish between two sorts of sensors: user-end sensors and ambient sensors. The first group comprises the 14 EEG sensors that were previously submitted and placed on the user's headgear. The second set of sensors is primarily responsible for allowing the arm to connect to and adapt to its surroundings by providing intelligent input in critical situations such as high temperatures or pressure. This network would give the prosthetic arm a human-like action with cerebral reactions and smooth motions when connected to the embedded microprocessor mounted on the arm [13,15]. The data from each of these sensors will be displayed on a small LCD screen on the forearm and will be used to control individual arm servos.

1.2.2.3 Electromyography

This is a technology that tracks and evaluates the electrical activity that skeletal muscles emit [16]. An electromyograph that generates an electromyogram is used to conduct electromyography (EMG). The instrument typically measures the electrical potential of muscle cells on a person's skin when the muscles are activated electrically or neurologically. It is then essential to analyze the signals. EMG is also used for gesture recognition, and identifying the triggered action in the prosthetic area enables the information to be transferred to the prosthetic hand to conduct it. EMG can be achieved and applied to the skin using electrodes, called surface EMG (sEMG). This can be achieved invasively by directly injecting electrodes into the muscles. This procedure is invasive and can induce exclusion or cause infections [17]. To deal with the signals and obtain concrete behavior, the signal recorded from sEMG must then be transmitted to a computer. For example, the latest Bluetooth and WiFi technology can be used on a mobile processing unit (e.g., a smartphone). Low voltage requirements, such as Bluetooth low energy and low power WiFi, can be used to increase battery longevity.

1.2.2.4 Flexors and extensors

The prosthetic has to rotate the hands using flexors and extensors to grip the piece. The flexor helps the hand to close. Many people use voluntary closing devices, which ensure that, where no force is involved, the normal position is free. The hand closes when triggered. To cover the sides, cords or nonelastic bands are used. Flexion of the wrist or elbow allows the wires or strings attached to the ends of the fingers to maintain a tight grip [18]. There is also the use of mechanical solutions without wires or cords. Engines or even compressed air can influence mechanical connections. The prostheses usually use elasticity to stretch the limb, which automatically opens the arm. This function is assured by elastic strings, bands, and finger joint elasticity. Elastic finger joints are conformant mechanisms. The prostheses have fingers constructed from one piece with rigid phalanges and adjustable joints using these compatible structures. Every prosthesis has motor-attached wires, chains, or mechanical connections.

1.2.2.5 Kinematic aspect

Based on their structure, the prosthetic hands may execute various gestures and grasps. The number of fingers, degrees of freedom, number of valves, range of movement, and forms of the grip of the joints all affect the prosthesis' kinematic parameters [18]. Owing to the relationship with the phalanges in the fingertips, most prosthetic hands have more degrees of freedom than actuators. As mentioned before, there are normally three phalanges connected via cables, cords, or some other wire. Each finger's cable is then linked to a common connection to ensure that the fingers travel at the same time [19]. The externally driven prostheses have motors attached to each finger that allow each finger to be operated independently. There are four joints in a human finger: the distal interphalangeal joint, the metacarpophalangeal joint, the carpometacarpal joint, and the proximal interphalangeal joint. The thumb itself has just three joints, replacing the distal interphalangeal joint with the interphalangeal joint. There is no carpometacarpal joint for the prosthetic paws.

1.2.2.6 Daily life grasps

To guarantee the essential tasks of everyday life, the prostheses should have various forms of grip. To conform to the shape of the items they are carrying, they will need to execute an adaptive grip. To guarantee that some fingers can exert force when most others are hindered, there might be an intensity ratio between the fingers [18]. They can be operated independently by prostheses that have motors on each finger. By adjusting the friction in the fingers using levers, an air pressure-actuated hand will exercise the grip. Some have an intelligent system that distributes the force.

1.3 Results

Only sEMG electrodes are presented in this chapter. There should be separate electrodes that can sense the electric field of the muscle. The Advancer Technologies business provides a fair price for sEMG sensors. The company manufactures a MyoWare Muscle Sensor that can be used to provide the required feedback for the prosthesis. The explored single-channel EMG boards allow microcontrollers, such as Arduino, to be easily used. These multiple EMG electrodes now produce acceleration captors that are used for improved accuracy and thus a greater activation of motion in comparison to the electrical signal of the muscle. The most widespread industrial electronic hand control method is quoting the electrical signal to the muscles' EMG, which has many advantages, but its use requires very complex circuits to process and adapt this signal. Filter circuits are essential in this application because the jamming signals are extracted from the sensors (signal switchers). Whether these signals are from within the human body or from outside, the control of the prosthetic hand requires multiple filtering circuits in addition to the full protection of the control circuit by elaborate and effective encapsulation [19]. This has led to numerous attempts at borrowing another voluntary signal from the amputated human body that would be more effective without requiring complex control circuits in terms of design and manufacturing. Simplifying the control circuit, thus reducing the size that this circuit occupies, helps facilitate the mechanical design of the prosthesis. This gives higher efficiency and better quality in all respects [16]. The proposed system controls the electronic prosthetic hand by quoting the signal of the mechanical muscle movement to an amputee, using a motion sensor or pressure sensor placed on the outer surface of leprosy.

1.3.1 3D printing arm

Individuals are born with specific skills. Due to illness or injury, they may lose these abilities. They may become unable to speak, see, or hear, and therefore lack a significant amount of knowledge and information about the world. As a result their response to the world becomes tougher, and they become more reluctant. Today, humanity strives to crack the boundaries of the biological body's capacities, add additional capabilities, and make up for other limitations, and not succumb to obstacles and difficulties.

The proposed system concentrates predominantly on people who have been victims of terrible injuries or paraplegia dysfunction that has caused the destruction or loss of their arm. The mechanical arms provided by the healthcare system worldwide are extremely high priced [1]. Rapid prototyping methods, especially 3D printing methods, may offer a versatile way of producing a bionic arm to act as a replacement [1]. There are more solutions to making a robot arm in this modern age of technology. As we dive deep into the technology and the production method of prosthetics, we also learn that the diverse methods used to manufacture prosthetic arms are very expensive. It is also not possible for a person belonging to a poor economic status to afford a prosthetic hand. The robotic limbs are commercially available and cost up to $20,000−$30,000. For this purpose, we have developed a less expensive device accessible to both the rich and the poor. Since the robotic arm is an active body part, we have synchronized the robot arm and the human body.

1.3.2 Solution and plan of the proposed problem

In earlier times, when human beings lost a body part, for example, fingers, feet, and so on, they used a timber or steel prosthetic piece to substitute the side. However, such a prosthetic piece becomes a rigid component of the body and would be difficult to be merged with other body parts because of its rigidity and rough edges. Another drawback was that it would generate resistance to the normal motion of the human body. Body parts made of timber or steel would serve no other purpose but to meet standards of beauty. As time progressed, this field of study experienced significant development, culminating in the invention of the robot arm. There has been a lot of progress in this area and, recently, prosthetic arms can perform much of the usual function of the human upper limbs. As physical, electrostatic, and environmental systems technologies have enhanced and developed, we have technological gadgets such as sensors that would allow the robotic arm to perform precise and required movements. Most of the existing prosthetic arms cost around

$20,000−$30,000, which makes them inaccessible to the poor. We capitalized on this to develop an inexpensive, practical human prosthetic arm.

1.3.2.1 Required materials

All fundamental principles and frameworks for the processing of the component were made available for additive manufacturing. The basic prerequisite of 3D product development is the establishment of a product design dependent on software. We intermittently used a separate CAD program to create a digital design: Hardcore, Creo, Matlab, Mixer, Asays, etc. We established certain specifications for production to produce the robot arm using the additive manufacturing printer. Several programs were used to establish this measurement, for example, bed temperature, cooling time, object orientation, subject dimensions, lattice constant, etc. One of the applications used by this service is Cura [18]. After setting all the program parameters, we fed the file created by the software into the 3D printer. The 3D printer scanned the document and generated the model based on the current dimensions and specified parameters.

The materials and methods used to manufacture the proposed device include:

1. The first piece is called the Pylon, which is the prosthesis brace or inner frame. It provides stability for the limb and is primarily made of metal links, although it has recently been used to make carbon fiber. This piece is protected by another material that, according to the skin tone of the party holder, can be assigned the chosen color to give the party a more natural appearance.

2. The other component is the socket in which the prosthesis is mounted, which is installed on the remaining part of the amputated limb and thereby moves the weight from the body of the recipient to the prosthesis. The socket must be installed in a manner that it removes repetitive contact with the body to prevent trauma to the skin and tissues underneath, injury, and infection. The patient should be given a form-fitting sock, and the prosthesis should be securely attached to it.

3. And the last part is the suspension device that serves to hold the prosthesis fixed to the body, and these processes vary depending on the prosthesis since various belts and sleeves are often used to connect the body to the arm.

In certain kinds of amputations the arm may remain attached to the body only because of the form of the remaining part of the limb, and often a locking device is introduced that enables the prosthesis to stay attached and locked in place. These specific components are shared among most prosthetics, and each model of these limbs varies due to its configuration and the form of amputation arising, whether above or below the joint, resulting in fundamental variations in the type of prosthesis needed. For example, amputation above the knee involves a form of prosthesis that has a prosthetic shoulder within it, whereas amputation below the shoulder joint relies on interaction with the shoulder joint.

1.3.3 Methodology of concept

The quality or great outcome of an innovation depends on how correctly and specifically the design is sampled. Designing every object involves some basic steps. The first phase is the perception of the plan, when it is important to form a clear understanding of the design objective. In the area of the bio-cad mechanical arm, we agreed to do research recognizing the need for disadvantaged patients who are not able to afford the costly robot arm. For this purpose, we created a robot arm that is affordable to both the poor and the rich. The outlook of the prosthetic limb research will be affected by the introduction of the principle of additive manufacturing in the world of biomedical technology. We needed a simple base to implement the additive production process. A new image of the object is required for the production of any component by additive manufacturing, so the working prototype is evident from that. The whiteboard for our mechanical arm had to be created. A lot of programming ideas and concepts were exhibited to create the computer algorithm of the mechanical arm. We primarily concentrated on an application called Blender. It is the most commonly used program to create the nervous system component model. But the design and the processor are required to be consistent with each other. We discovered from the literature surveys that CAD technology is most frequently used for composite materials. So, we eventually agreed to develop a CAD robot arm called the Creo. The Creo concept model is conveniently compatible with the explored 3D printing technologies and tools. We fulfilled all the necessary parameters of the mechanical arm when constructing it. Since our target was to create a human prosthetic arm, one of the most important design criteria is the style of the prosthetic arm. Health is also a critical architectural criterion in the robotic arm style [5]. We also aimed to prevent system collapse. The designed robotic arm has been tested and confirmed to fulfill the needs of the patients. We utilized a node capture method called Ansys for analysis purposes, and the performance was outstanding compared with the existing devices.

1.3.4 Product development

The greatest difference between old and new prosthetics is the modern materials (made of carbon fibers) used in their manufacturing, which offer the prosthesis lightness, strength, and a natural form. As for electronics, it helps to provide more control over the prosthesis for multitasking, collecting, or walking on various objects. The framework for the manufacturing process establishes a connection between the device's history, present, and future [4]. It includes a general idea of the product, its characteristics, and the functionality of the system. The main aim of the device is to have a cost-effective human prosthetic arm, so that any individual who requires this service can easily afford it. The product used for producing the implant's arm is a different composite component, so the shape of the implant hand is much smaller relative to the actual implant hand. The other cool and cheerful-shaped robotic arm is mainly enjoyed by children. Compared with the new mechanical arm, the price of the service is much lower, so any person can comfortably afford it [17]. Although, in most situations, a prosthetic hand is already built for the infant, we have to remodel the robotic hand according to the proportions of the arm as the child grows. So, if we consider the expense factor of it, then relative to the 3D-printed mechanical arm, it would be too costly. This arm is rather robust, and with it, you can do most of the functions that the normal hand does, from opening doors to closing the fist, among others. Special gloves can be worn to hide these limbs and make them resemble a human hand.

The design of the prosthesis system should be able to mimic human movements as closely as possible. The absence of dexterity in the prosthetic arm induces compensatory movement that can cause long-term injuries. Some problems still need to be solved, but certain prosthetic arms are working. Reasonable actuators and mechanisms for moving the arm are included in the design of the upper limb prosthesis. The anthropomorphic shape limits the size of actuators whose high-power density is required for the prosthesis to exert the torque and force necessary for action during service. The control system helps the patient to use the prosthesis device correctly along with the sensors [14]. The different components of the prosthetic arm include the user input, control system, power source, actuators, sensors, and mechanical structure.

As a spherical parallel manipulator, the shoulder of the prosthetic arm was shaped like that developed by Gosselin for other purposes. This manipulator consists of three identical arms. Each arm consists of three revolutes arranged to converge each rotation axis to the same rotation center. The manipulator is free of rotation of 3 degrees in this arrangement. These DOFs permit the motions of shoulder abduction-adduction, flexion-extension, and inter-external rotation. The flexion-extension elbow motion is performed using a four-bar system. This allows the actuating engine to be mounted next to the leg. The closer the weight is to the shoulder, the lower the rotational inertia. In addition, the four-bar link can be set when the motor delivers higher torque where it is most needed [7]. The wrist is formed by a spherical manipulator that facilitates the movements of forearm pronation-supination, ulnar-radial deviation, and wrist flexion-extension. The proposed system is assembled to form an anthropomorphic prosthetic arm. The shoulder requires, with its actuators, the 3 DOF spherical manipulator. Using a link, the end effector of the manipulator is rigidly attached to the forearm structure. At the base of the forearm, the motor that actuates the elbow is fixed. The spherical mechanism facilitates the three movements that the shoulder requires, and the three motors share the load with tiny motors. The prosthetic arm can therefore be built with anthropometric form and volume. The focus of this work is on the design of the high limb without a hand, but for simulation purposes, a mass of 500 grams represents the hand load. The total weight of the design is 1,250 gr, including the side. There is enough space for the supply and control system inside the structure of the prosthesis [6].

1.3.4.1 Motion simulation and results

Solid Works has been modeled on the proposed design [13] to test the actions of the system following three different trajectories on which a dynamic simulation was carried out. To construct the actuation system, the output torques were acquired. In all the simulations, a load of 10 N was added to the hand to reflect the payload that the prosthesis would be able to pass [7]. Here:

1. The shoulder flexion motion was the first activity simulated. The amplitude of the motion, beginning with the fully vertical arm, was intended to be 90 degrees. The motion length with a sinusoidal velocity was 2 seconds.

2. An abduction motion with the prosthetic placed in the horizontal plane was the second simulation motion. The motion amplitude was 70 degrees with a period of 2 seconds. The time simulation consisted of flexing the elbow from 0 degrees to 100 degrees for a duration of 3 seconds.

For the first and second simulations, the results show that the three motors perform smooth motions when the prosthesis follows the necessary trajectories [13]. The maximum torque happens at a time different from the maximum velocity. It can be shown that at the end of the movement, the highest torques for motors 1 and 3 occur, while motor 2 produces the greatest torque at t = 1.4 seconds. But it has a maximum torque of 6Nm. For the second test, motor 1 is the key component of the action. At the very beginning of the simulation, the highest torques occur for engines 1 and 2. At the final stage of the motion, engine 3 unexpectedly rises to its full value at the end of the action. The torque limit is 15Nm. The motor conducts a smooth sinusoidal movement for elbow flexion, with a small torque variance occurring at the center of the motion concerning the maximum torque of 6.2 Nm [6].

1.3.4.2 How long does the prosthetic arm last?

The prosthesis can last anywhere from several months to several years, depending on the patient's age, activity level, and development. Many changes in the residual limb occur in the early stages after limb loss, which may contribute to limb shrinkage. Socket adjustments, liners, or even a separate system can be needed for this. A need for a change in the prosthesis or its parts may be generated by an increased level of activity and a willingness to perform more activities. The prosthesis requires only minor repairs or maintenance once you are satisfied with the fit of your device, and can last an average of three years. To prevent any major issues, your prosthesis should be periodically tested by your prosthetist [4,20].

1.4 Conclusion and recommendations

The proposed arm hosts state-of-the-art technical innovation, procedures of correspondence, mechanisms of control, and human contact. In certain fields, whether relevant to the area of healthcare or not, this provides immense solutions. The term may be extended to various sections of healthcare services as well as to people with other pathologies, such as with nerve injury. Many manufacturers and consumers use many aspects of the suggested arm. There is a subset of people within the healthcare world who need additional assistance in their everyday lives. It covers disabled people, rehabilitated adults, people with restricted movement, etc. The proposed arm could be attached to a robot system and presented to these individuals as assistants or caregivers. Similar to particular patient specifications, it can be configured to perform different tasks. This can range from preparing to washing or clothing assistance. In a high-workload surgical technique, where doctors perform procedures manually with the assistance of a prosthetic hand, is another example of its use in the medical industry. In the development phase, many companies employing robotics can make use of a tweaked version of the proposed arm.

Controlling prosthetic hands using the muscle movement method is one of the modern technologies that can be combined with various other modern technologies. Here, the sensor can be replaced by an array of sensors located on the whole or most of the rhizome area, and then we train a neural network so that the output of the sensors is converted into control signals for the movement of each finger. To do this, a group of muscles contracted as a result of their movement before amputation, moving the index finger, and by using neural networks, we associated each group of readings with a motor control signal connected to the finger parallel to the finger that causes this group of readings (i.e., what we asked the patient to imagine moving). Here, we have been given the advantage of moving each finger separately, but it is also possible to use the control signal that we got to control the rotation of the wrist prosthesis, or any other movement that may be important to the patient. And if we use a programmatic neural network, the control signal will be stored on a microcontroller, and therefore various programs can be placed on this controller that would enable the patient to control things other than the hand, for example, various finger movements can be stored, and a program can be used to control a mobile phone, a radio, or even a remote control for a TV or a CD playback device, without interfering with the main work of the sensor network, where the patient chooses what he controls by visualizing the movements of the fingers. This technique may be very useful for amputees with both forearms.

We manufactured the final form of the circuit of the proposed device in an integrated manner (i.e., in a single integrated circuit), and after achieving this, we got rid of all the resistors and capacitors. Placing piezoelectric pressure sensors on the prosthetic fingertips and connecting them in the form of a feedback circuit, gives the patient a greater ability to control the pressure of the artificial handpiece (engine torque).

References

[1] F. Cordella, et al., Literature review on needs of upper limb prosthesis users, Frontiers in Neuroscience 10 (2016) 209.

[2] Y. Li, M. Alam, S. Guo, K.H. Ting, J. He, Electronic bypass of spinal lesions: activation of lower motor neurons directly driven by cortical neural signals, Journal of Neuroengineering and Rehabilitation 11 (1) (2014) 1−12.

[3] D.K. Kumar, B. Jelfs, X. Sui, S.P. Arjunan, Prosthetic hand control: a multidisciplinary review to identify strengths, shortcomings, and the future, Biomedical Signal Processing and Control 53 (2019) 101588.

[4] Y. Wang, Y. Tian, H. She, Y. Jiang, H. Yokoi, Y. Liu, Design of an effective prosthetic hand system for adaptive grasping with the control of myoelectric pattern recognition approach, Micromachines 13 (2) (2022) 219.

[5] J. Andrés-Esperanza, J.L. Iserte-Vilar, I. Llop-Harillo, A. Pérez-González, Affordable 3D-printed tendon prosthetic hands: expectations and benchmarking questioned, Engineering Science and Technology, an International Journal 31 (2022) 101053.

[6] S.P.S. Yadav, V.K. Shankar, L. Avinash, A. Buradi, B.A. Praveena, V.K. Vasu, et al., Development of 3D printed electromyography controlled bionic arm, Sustainable Machining Strategies for Better Performance, Springer, Singapore, 2022, pp. 11−21.

[7] J.E. Uellendahl, Upper extremity myoelectric prosthetics, Physical Medicine and Rehabilitation Clinics of North America 11 (3) (2000) 639−652.

[8] U. Wijk, I.K. Carlsson, C. Antfolk, A. Björkman, B. Rosén, Sensory feedback in hand prostheses: a prospective study of everyday use, Frontiers in Neuroscience 14 (2020) 663.

[9] A. Arabian, D. Varotsis, C. McDonnell, E. Meeks, Global social acceptance of prosthetic devices, in: 2016 IEEE Global Humanitarian Technology Conference (GHTC), IEEE, Seattle, WA, 2016, pp. 563−568.

[10] L.E. Osborn, J.L. Betthauser, N.V. Thakor, Neural prostheses, Wiley Encyclopedia of Electrical and Electronics Engineering (1999) 1−20.

[11] L. Barton, M. Chavez, What is the effectiveness of 3D-printed orthoses for people with upper extremity amputations in improving the performance of daily activities? 2021.

[12] M.C. Carrozza, F. Vecchi, F. Sebastiani, G. Cappiello, S. Roccella, M. Zecca, et al., Experimental analysis of an innovative prosthetic hand with proprioceptive sensors, 2003 IEEE International Conference on Robotics and Automation (Cat. No. 03CH37422), 2, IEEE, 2003, pp. 2230−2235.

[13] C. Ahmadizadeh, L.-K. Merhi, B. Pousett, S. Sangha, C. Menon, Toward intuitive prosthetic control: solving common issues using force myography, surface electromyography, and pattern recognition in a pilot case study, IEEE RobotIcs & Automation Magazine 24 (2017) 102−111.

[14] C. Cipriani, M. Controzzi, M.C. Carrozza, Objectives, criteria and methods for the design of the SmartHand transradial prosthesis, Robotica 28 (2009) 919−927.

[15] C. Cipriani, M. Controzzi, M.C. Carrozza, The SmartHand transradial prosthesis, Journal of Neuroengineering and Rehabilitation 8 (2011) 29.

[16] M. Bridges, J. Beaty, F. Tenore, M. Para, M. Mashner, V. Aggarwal, et al., Revolutionizing prosthetics 2009: dexterous control of an upper-limb neuroprosthesis, Johns Hopkins APL Technical Digest 28 (2010) 210−211.

[17] https://www.medicaldevice-network.com/news/prosthetic-arm-gives-wearers-a-sense-of-touch/ Retrieved August 17, 2021.

[18] https://www.researchgate.net/publication/345726775_Availability_of_sEMG_controlled_prosthetic_arm_components Retrieved May 9, 2021.

[19] C. Lake, J.M. Miguelez, Evolution of microprocessor-based control systems in upper extremity prosthetics, Technology and Disability 15 (2003) 63−71.

[20] J.L. Collinger, B. Wodlinger, J.E. Downey, W. Wang, E.C. Tyler-Kabara, D.J. Weber, et al., High-performance neuroprosthetic control by an individual with tetraplegia, Lancet 381 (2013) 557−564.

CHAPTER TWO

Internet of things-based patient well-being monitoring system

Dilber Uzun Ozsahin[1,2,3], Basil Bartholomew Duwa[3,4], John Bush Idoko[5], Angelique Rwiyereka[4], Deborah Ishimwe[4], Shahida Hassan[4] and Ilker Ozsahin[3,6]

[1]Department of Medical Diagnostic Imaging, College of Health Science, University of Sharjah, Sharjah, United Arab Emirates
[2]Research Institute for Medical and Health Sciences, University of Sharjah, Sharjah, United Arab Emirates
[3]Operational Research Center in Healthcare, Near East University, Nicosia/TRNC, Mersin 10, Turkey
[4]Department of Biomedical Engineering, Near East University, Nicosia/TRNC, Mersin 10, Turkey
[5]Applied Artificial Intelligence Research Center, Department of Computer Engineering, Near East University, Nicosia/TRNC, Mersin 10, Turkey
[6]Department of Radiology, Brain Health Imaging Institute, Weill Cornell Medicine, New York, NY, United States

Contents

Practical Design and Applications of Medical Devices.
DOI: https://doi.org/10.1016/B978-0-443-14133-1.00009-4
23

2.1 Introduction

While a medical institution is, without a doubt, a safe place for patients, nurses who manually carry out tests/analyses on the symptoms of diseases ought to be monitored. Uncontrolled shifts among nurses can lead to a patient's death [1,2]. However, the Internet of Things (IoT) can be a platform that lets in for interplay to exist among considerable gadgets, vehicles, buildings, and exceptional materials capsulated with electronics, software, sensors, actuators, and community assets that allows those gadgets to build up and exchange information over the community of networks known as the Internet. However, those embedded IoT sensors use various types of community connections similar to WiFi, Bluetooth, etc., and have been carried out in numerous regions along with environmental tracking, infrastructural management, delivery, and manufacturing.

The predominant rudiments of the wearable IoT era are the sensors and gadgets, which are significantly energy-efficient, have low energy intake, and are petite-sized. In this study, we propose a gadget based on IoT for patients' fitness. In general, patient data includes patients coming to/from hospitals, clinics, and exceptional fitness centers to be tested by clinical employees. On arrival at the fitness facility, patients pay a considerable amount of money at the reception to enable them to see the medical practitioners and different allied fitness specialists. When a patient is admitted to the medical institution, monitoring is done intently to visualize the remedies administered to them by the specialists. However, there are two sorts of patients whose well-being is monitored, namely incoming patients and outgoing patients. Incoming patients are patients who visit hospitals for healthcare check-ups for analysis and treatment of ailments [3]. These patients are from time to time admitted into a ward after examination based mostly on the recommendation made by the medical doctor on duty. It becomes further stressful as patient health monitoring is done at precise levels in many nations' health areas, namely primary healthcare (local government level), secondary healthcare (state government level), tertiary healthcare (federal government level), and private healthcare (private practice level).

This study presents a novel interconnected gadget capable of identifying critical symptoms such as temperature and heartbeat rate. The tracking gadget turned into an IoT platform is made up of Arduino Mega 2560 and ESP8266 Wi-Fi Module. Two sensor modules are accustomed to each critical signal stage, where every module uses a temperature sensor [4]. The major goal of this research is to build a patient tracking gadget capable of checking critical symptoms and symptom stages, examining the severity of vital symptoms in line with the patient's age, providing alerts for abnormalities, and simultaneously displaying the effects wirelessly through humanoid apps. This venture will limit the workload of nurses in medical institutions and deliver many handy techniques in the statement status of each symptom for patients inside the ward.

The standard approach of nurses visiting each patient to report vital symptoms time-consuming [5]. With this technique, nurses can display patient information through humanoid apps installed on any Android device. Nurses or doctors might also evaluate the preceding critical signal details by just downloading the statistics from the cloud in the standard format. Comparison of the critical symptoms and symptom stage acquired from this gadget with everyday size instrumentation or a guide statement has proven sincerely comparable effects. One of the challenging issues faced by hospitals each day is patient tracking [6,7]. With the IoT-based gadget, the total fitness of a patient can be monitored and shaped anywhere in the world. The gadget presents specialists with critical statistics via electronic mail or SMS to appropriately control a patient's fitness.

2.1.1 Description of the proposed device

Health tracking is an important problem in today's world. Due to people's strong desire for proactive health monitoring, individuals often face real fitness-related challenges. Currently, there are components of IoT gadgets that display patients' data on the screen over the web. Health professionals are also taking benefit of those sensible devices to keep a watch on their patients. IoT is fast revolutionizing the healthcare industry with the latest healthcare innovation start-ups [8]. In this study, we have constructed an IoT primarily based on a health monitoring system where statistics of patients' coronary heartbeat and frame temperature conjointly send an email/SMS warning that readings are going beyond primary values. Heartbeat rate and frame temperature readings are recorded over ThingSpeak and Google Sheets, so that the patient's data may be accessible over the web.

In this research, unique parameters of the patients are checked using net factors. In the patient tracking device primarily designed based on web factors, the real-time parameters of the patient's health are dispatched to the cloud using a web network. These parameters are dispatched to a faraway website so that the concerned individuals can see the information from anywhere in the world. There can be a brief comparison between SMS primarily based on the patient's health checking, and IoT, primarily based on the health practitioner. In IoT devices, information about the patient's health may be visible with the aid of several clients [9].

In this survey, all the systems have been ranked based on the need of the components, that is, which components have been utilized more than others. Hence, the IoT-based systems have been categorized into three distinctive categories: sensor-based wellbeing observing systems, smartphone-based wellbeing observing systems, and microcontroller-based wellbeing checking systems. Because of the increasing ubiquity of IoT and widened research openings, several architectural systems have been proposed for the usage of eHealth. Here, the primary goal is to lay out an IoT patient health monitoring system to diagnose the well-being situation of the outpatients. Giving care and health assistance to bedridden patients at critical stages with advanced medical facilities have become major problems in the modern world. Proper implementation of such structures can offer well timed warnings to the scientific staff and medical doctors, and treatment may be activated in cases of clinical emergencies [10]. The use of sensors detects the state of affairs of patients, and the statistics is amassed and transferred with the help of a microcontroller. Doctors and nurses may not be able to go to the patient often to take a look at their current condition, hence the need for the proposed system.

2.1.2 Usefulness of the proposed system

1. IoT monitoring system demonstrates supportive features when programmed to monitor the fitness parameters of the patient over a given period of time.
2. Clinical facilities are minimized because of remote patient monitoring.
3. Clinical visits for normal habitual check-ups are minimized.
4. Patient wellness parameter statistics is placed over the cloud. So, it is more beneficial than preserving the statistics on published papers. Or certainly the virtual statistics that are stored in a selected PC, or pill, or reminiscence tool like a write-drive.

2.2 Components and working principles of the proposed system

2.2.1 Components

This section presents the components used in the construction of the proposed device as well as their working principles. Table 2.1 depicts the names and descriptions of the implored components

2.2.2 Working principles

The IoT-based patient monitoring system has two sensors: the temperature sensor and the heartbeat sensor. This undertaking could be very beneficial because the medical doctor can screen patients' fitness parameters simply by surfing an Internet site or URL. Here, we constructed an IoT-based health checking system that depicts the statistics of patients' coronary heartbeat rate and frame temperature and sends an email/SMS caution at any factor beyond the basic/recommended values. Through ThingSpeak and Google Sheets, pulse rate and temperature measurements are registered so that records of a given patient may be found anywhere on the Internet worldwide. A warning notification is also incorporated so that patients can click on it to send email/SMS to their own circle of relatives in an emergency [11]. To use the proposed novel system, you need a WiFi connection for easy communication between the patient and the

Table 2.1 Components and details.

Sr. No	Materials	Description	Quantity
1	Arduino board and programming cable	Arduino Uno	1
2	Pulse sensor	Pulse heart rate sensor	1
3	Temperature sensor	LM35 analog temperature sensor	1
4	ESP8266	ESP8266−01 WiFi module	1
5	LCD		1
6	Potentiometer	10 k	1
7	Resistor	2 k	10
8	Push button	Push button	1
9	Connecting wires	Jumper wires	10−20
10	Breadboard	Medium	1
11	LED	5 mm LED, any color	2

relatives/friends of the patient. Using a WiFi module, the microcontroller or the Arduino board connects to the WiFi network. You can create a WiFi zone using a WiFi module, or else you can activate a WiFi zone on your mobile device using a hotspot. The Arduino UNO board continuously analyzes feedback from these two sensors; it sends this information to the cloud at any point by sending it to a particular URL/IP address. At such a point, after a specific interim period of time, this operation of sending data to IP is rehashed [7].

2.2.2.1 Arduino Uno

The Uno is the simplest, most used, and also the most documented of the Arduino microprocessors. This type of microcontroller is most suitable for small projects, and beginners will face no problems working with it. It can be used by scholars for projects in different fields. The Uno is equipped with all the functions needed to use it with great efficiency [7]. It has a USB port that allows the microcontroller to be connected directly to the PC, which is also an interlink that plays the role of transferring code (programming) to the microprocessor, which is achieved with the aid of an integrated development environment (IDE). Fig. 2.1 depicts the frontal view of an Arduino Uno.

2.2.2.2 Pulse rate sensor

Electrocardiography remains the best choice for measuring blood pressure because the pulse rate sensor is simple and convenient. It is cheap and can be used in small wearable devices, such as watches and smartphones. It

Figure 2.1 The Arduino Uno.

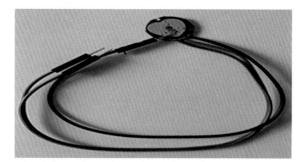

Figure 2.2 The pulse rate sensor.

can integrate well with Arduino. By putting the pulse sensor in contact with the ear or fingertip and connecting it to the Arduino, the heartbeat rate can be easily measured [2]. The pulse rate sensor is shown in Fig. 2.2.

The pulse rate sensor is made up of a diode that gives out light and a photodiode (resistor) that discerns light on one surface. A noise reduction and amplifying circuit is attached on the other surface. The blood flow is measured by means of an LED light that is placed on the fingertip, where a vein should be present. As the blood flows, the LED emits light and the light sensor begins to receive light due to the flow of blood, and thus, by analyzing the difference in light gained, we can know the pulse rate [1]. The proposed system is useful for monitoring anxiety and for remote monitoring of patients. It can also be used in gaming consoles as well as in health bands.

2.2.2.3 Temperature sensor

Apart from the thermometer, which is a well-known instrument for measuring temperature, temperature sensors can be designed for low-cost projects. One such sensor is the LM35, designed to work well with Arduino microcontrollers or any other microcontrollers with the same features. It is also called the numerical thermometer. It can sense temperatures not lower than -50°C and not higher than 150°C, which works well for our project. It can be interlinked with the Arduino Uno with great ease. The L35 is a tri-pin sensor that, for every 10 mV difference, produces 1°C; therefore the output voltage is directly proportional to the temperature [3]. The working principle of the LM35 temperature sensor can be divided into three stages: First, the sensor senses the temperature; second, the whole scaling process of the conversion of the temperature previously

Figure 2.3 The LM35 sensor.

sensed is done by Arduino; and finally, the temperature is shown on the LCD. Fig. 2.3 shows the LM35 sensor.

2.2.2.4 ESP8266

The ESP8266 is an exceptionally user-friendly and low-cost gadget that supplies a web network to your proposed system. The module can work both as an access point (can make a hotspot) and as a station (can interface with WiFi). Thus it can effortlessly get information and transfer it to the web, making the Web of Things as simple as conceivable. It can get information from the Internet utilizing API (application programming interface), which subsequently allows access to any data that is accessible on the web [8]. Another amazing highlight of this module is that it can be modified utilizing the Arduino IDE, which makes it a package more client-friendly. Fig. 2.4 depicts the ESP8266.

2.2.2.5 LCD

For the result read-out interface, we used the LCD to display status messages or sensor readings. LCD presentation is the ideal fit. It is an incredibly normal and quick method to add a comprehensible interface to the proposed system. An LCD is a level board display or an electronically tweaked optical gadget that utilizes the light-balancing properties of fluid precious stones joined with polarizers. Fluid precious stones do not discharge light directly,

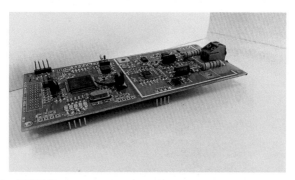

Figure 2.4 The ESP8266.

Figure 2.5 The LCD unit.

rather utilize a backdrop illumination or reflector to deliver pictures in shading or monochrome. These LCDs are ideal for showing text/characters, hence the name Character LCD [12]. The presentation has a LED backdrop illumination and can show 32 ASCII characters in two columns, with 16 characters on each line. Fig. 2.5 shows the LCD.

2.2.2.6 Potentiometer

For this research, we utilized a 10k potentiometer. A potentiometer could be a three-terminal electrical device with a pivoting contact that shapes a customizable resistance. If just two terminals are utilized, one end and the wiper, it works as a variable resistor [10]. A typical potentiometer is displayed in Fig. 2.6 while Fig. 2.7 demonstrates the block diagram of the proposed system

2.2.3 Applications of IoT platform (ThingSpeak)

1. Uses the ThinkTalk platform to send data to the cloud from any device that is allowed on the Internet.

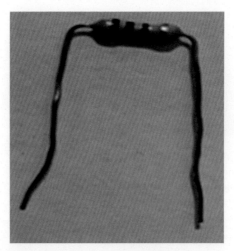

Figure 2.6 Potentiometer.

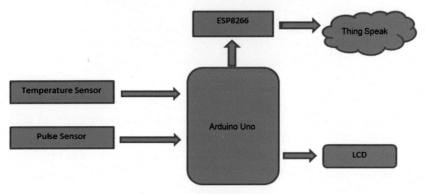

Figure 2.7 Block diagram of the proposed system.

2. Actions and alerts can then be configured based on real-time information and unlocked by means of visual tools information.
3. Use ThinkTalk to provide a platform for developers to quickly collect data from sensors and translate it into usable data.
4. The ThinkTalk server is an open data forum and is an IoT API that allows you to capture, store, study, visualize, and respond to data from sensors.
5. ThingSpeak is available for noncommercial small businesses as a free benefit ($<$ 3 million messages annually and $\sim$8200 messages every day). Tools for ThingSpeak are manufactured in units where 333 units are allowed in one unit. Millions of messages that should be prepared and saved in a one-year cycle ($\sim$90,000 messages every day).

2.2.4 Connections with LCD

1. Connect the A0 output pin to the Arduino pulse sensor and the other two to the VCC (voltage common collector) and GND (ground) pins.
2. Link the LM35 temperature sensor output pin to Arduino's A1 and the other two to the VCC and GND pins.
3. Link the LED to the Arduino digital pin 7 via the Ohm Resistor 220.
4. Link the LCD pin to GND 1,3,5,16.
5. Link the 2.15 LCD pin to VCC.
6. Link the LCD pin to 4,6,11,12,13,14 for Digital, Arduino, pin 12,11,5,4,3,2.
7. The RX pin of ESP8266 operates at 3.3 V and will not work at 3.3 V.
8. Link the LCD to the Arduino when we want to connect it directly to the Arduino. A voltage divider will then be created to convert from 5 V to 3.3 V. This can be achieved by linking resistors of 2.2 K and 1 K. The RX pin of the ESP8266 is then attached to pin 10 of Arduino by means of the surrounding resistors.
9. Link the ESP8266 TX pin to Arduino pin 9.

2.2.5 Connections without LCD

1. Pulse sensor signal pin –> A0 for Arduino
2. Pulse sensor VCC pin –> 5 V of Arduino
3. Pulse sensor GND pin –> GND of Arduino
4. LM35 Vout –> A1 of Arduino
5. Tx of ESP8266 –> Arduino pin 10
6. Rx of ESP8266 –> Arduino pin 11
7. ESP8266 CH PD and VCC –> 3.3 V Arduino
8. GND of ESP8266 –> Arduino GND of ESP8266
9. Push button –> Arduino digital pin 8

2.3 Implementation

ThingSpeak provides an incredibly fun instrument for IoT-based projects. We can track our information and manage our system over the Web through the networks and web pages by using the ThingSpeak site. ThingSpeak collects, analyzes, and visualizes sensor data, and acts by triggering a response. ThingSpeak is used to track the pulse and temperature of patients online through the Internet.

To start working with ThingSpeak, you have to sign up. To do that, visit https://thingspeak.com and create an account. Create a modern channel at that point and customize preferences. Build the API keys at that point. For programming alterations and setting your data, this key is needed. Move the code to the Arduino Uno at that point by accepting the circuit that appears above. Open the serial screen to naturally interface and set up the environment with WiFi. Press on channels now so that you can simply see the streaming of online information, i.e., heartbeat rate observation over the Web using ThingSpeak and ESP8266 with Arduino.

2.3.1 ThingSpeak to track information and manage system

1. First of all the customer must create an account on ThingSpeak.com, then sign in and click Get Started.
2. Go to the Channels menu and click on the Modern Channel alternative for further processing on the same page.
3. You will see a shape for making the channel at present; fill in the title and depiction as per your choice. In Field 1, Field 2, and Field 3 names, fill in "Pulse Rate", "Temperature", and "Panic" and tick the Area checkboxes. Underneath the shape, tick the check box for "Make Public" choice, and finally, Spare the Channel. Your unused channel is currently being created.
4. You'll see the three charts below as they appear. Note that we will use the Compose API key in our code. This key is required for programming alterations and setting up your data.
5. In our setup, we utilized the ThingHTTP app of the server to trigger the IFTTT applet for information passage to Google Sheets and send email/SMS. ThingHTTP enables communication among gadgets, websites, and web administrations without having to execute the protocol at the gadget level. It is possible to indicate activities in Thing HTTP, needed to trigger utilizing other ThingSpeak apps, such as React. To make Unused Thing HTTP, a URL is required for activation, often obtained from IFTTT.

2.3.2 How to configure IFTTT to trigger mail/SMS based on ThingSpeak values

1. Log in to IFTTT and tap on Webhooks.
2. In the Press documentation, there is a link to be used later.

3. Inside the event box, sort "Patient Info" and repeat the URL. This URL will be used in Thing HTTP. Now render ThingHTTP to Google Sheet guide Applet and send email/SMS. After that, jump into our ThingHTTP absolute.

4. In the My Applet option, click New Applet.

5. Click " + this" and check and tap on it for Webhooks. Select the "Receive a Web Request" trigger.

6. Sort the title of the occasion that is the same as that you wrote in the URL of Webhooks inside the occasion box. Click the Trigger button.

7. Press " + that" and hunt for and click on Google Sheets. Tap on the Include Spreadsheet drive.

8. Allow your sheet to include any word. You have the date and time, occasion title, BPM appreciation, and body temperature in the built push box that can be composed.

9. Survey the applet and press the end button. In the same way, we had to make an applet for sending emails when "Panic" event occurred. Tap " + this" again, and choose Webhooks, and enter "Panic". Browse for Gmail in " + that" and click on it.

10. Now, press the Email Send button. Enter the email addresses you would like to receive while the patient is in a panic state. In the Email, type the body substance you want to submit and click on activity creation. Study and finish this. To execute the tasks, we set our applets to Return to Thingspeak- > Apps- > ThingHTTP now. Fig. 2.8 depicts the complete setup of the proposed system.

Figure 2.8 Setup of the proposed system.

2.4 System software performance

2.4.1 Pulse measurement over ThingSpeak

As depicted in Fig. 2.8, throughout this research, using ThingSpeak and ESP8266 with Arduino, we performed pulse rate monitoring over the Internet, which detects the pulse rate using the pulse sensor and shows the heart rate (beats per minute) readings on the LCD as well as on the Internet. It sends the findings to the ThingSpeak server via the ESP8266 WiFi module, which helps to track the pulse from any part of the world via the Internet [11]. ThingSpeak is a nonproprietary IoT application and API for storing and retrieving information from subjects over the Internet or via a local area network using the HTTP protocol. The overall performance of the system is rated above 90%, as testified by the professionals.

2.4.2 Page for vital monitoring of the patient

Upon successful login by the doctor or the caretaker, either may view the vital essentials of the real patient's details regarding temperature, pulse, ECG, etc. With a view to securing the patient's data, the data is encrypted as it is sent to the ThingSpeak database server and is decrypted on the web page for data analysis.

The patient's readings are displayed on the critical monitoring page without any errors. In the event that the system is not connected or either sensor is not connected to the patient, all the readings or the corresponding reading will be displayed as zero in the case of digital values [9]. If the system is turned off, this page will only show the last known readings that have been stored in the database.

2.4.3 Testing the software

Software testing is a process in which a program is executed to detect bugs in the application. It is used to verify whether the application meets its standards and all the specifications. The application's functionality is practical. The ultimate aim of the research is to verify whether the test is under defined conditions, and whether the application behaves as envisioned. Every step to verify the accuracy of the application was examined and was successful. The primary objective of the research is to detect software flaws so that it is fixed. Cases of the test were built in full scope to detect bugs.

2.4.4 Testing categories

Testing can be conducted at many different categories, some of which may include the following:

1. Performance: It is the method of evaluating the speed of the system. This test is performed to ascertain the efficiency of a machine, network, software program, or unit, over a short period of time, and a lot of information is processed.
2. Testing of the unit: Testing of the unit applies to tests performed on a code segment to verify the piece of code's functionality. This is achieved at the level of operation.
3. Testing the system: A fully integrated system is evaluated by system testing to check that the precondition satisfies its specifications.
4. Assimilation testing: The implementation phase would involve any type of software testing that is intended to be performed to authenticate the component interfaces against a software design. Its key goal is to reveal the defects associated with the interface between modules.
5. Acceptance assessment: This assessment tests the compliance and diligence of the applications in meeting the conditions and specification criteria.

These tests were performed, and the proposed system recorded huge success at all stages.

2.5 Conclusion

This chapter presents a novel idea that would allow patients to monitor their vitals (temperature and heart rate) at any time. The system can send the health information over the web so that those concerned with the patient's well-being will be informed should any emergency occur, and the patient can receive the help they need at the appropriate time. The device modules could be incorporated into one optimized circuit. With the availability of the Internet around the globe, we can look forward to a future that prompts individuals to utilize such devices as the proposed device based on the health monitoring system to stay informed about their health in general. The device is cost-effective; which makes it easy for everyone to own and use it.

In spite of the fact that the IoT can be of an awesome advantage to healthcare, there are still major challenges to address before full-scale execution. Some of these challenges and drawbacks will be addressed in our future work, which include:

Security and privacy: Security and privacy are major concerns hindering clients from utilizing IoT innovation for therapeutic purposes, as healthcare layouts have the potential to be breached or hacked.

Risk of failure: Bugs within the equipment can affect the execution of sensors and associated gears, putting healthcare operations at risk. Skipping a planned program overhaul may indeed be more dangerous than skipping a specialist check-up.

Integration: There Is no agreement with respect to IoT conventions and guidelines, so gadgets created by diverse manufacturers may not work well together. The absence of conformity stops the full-scale integration of the IoT, thus restricting its potential efficacy.

Cost: While IoT promises to decrease the cost of healthcare with time, the cost of its application in health centers and staff training is high.

References

[1] A. Agnihotri, Human body respiration measurement using digital temperature sensor with I2c interface, International Journal of Scientific Research Publications (IJSRP) 3 (2013) 1−8.
[2] https://www.researchgate.net/publication/322542647_Iot_Patient_Health_Monitoring_System Retrieved June 6, 2021.
[3] https://www.researchgate.net/publication/316272664_Remote_Health_monitoring_using_Android_App Retrieved June 19, 2021.
[4] J. Holler, V. Tsiatsis, C. Mulligan, S. Avesand, S. Karnouskos, D. Boyle, From MachineTo-Machine to the Internet of Things, Elsevier, Amsterdam, 2014.
[5] A. Basra, B. Mukhopadhayay, S. Kar. Temperature sensor based ultra-low-cost respiration monitoring system, in: 2017 9th International Conference on Commununication and System Networks, 2017, pp. 530−535.
[6] https://circuitdigest.com/microcontroller-projects/iot-based-patient-monitoring-system-using-esp8266-and-arduino. Retrieved May 20, 2021.
[7] https://www.projectsof8051.com/arduino-and-iot-based-patient-health-monitoring-system-project/ Retrieved April 11, 2021.
[8] https://www.imaginis.com/heart-disease-diagnosis/heart-disease-monitoring-patients-with-cardiovascular-disease Retrieved June 4, 2021.
[9] A. Raji, P.K. Devi, P.G. Jeyaseeli, N. Balaganesh, Respiratory monitoring system for asthma patients based on IoT, in: Proceedings of 2016 Online International Conference on Green Engineering and Technologies (IC-GET), 2016, pp. 1−6.
[10] https://www.irjet.net/archives/V7/i3/IRJET-V7I3728.pdf Retrieved July 1, 2021.
[11] https://lastminuteengineers.com/arduino-1602-character-lcd-tutorial/ Retrieved July 1, 2021.
[12] https://www.google.com/url?sa = t&source = web&rct = j&url = http://www.jcre-view.com/fulltext/197-1585663661.pdf&ved = 2ahUKEwjvlODM39HsAhWIlYs KHT7VBhoQFjAPegQIHxAB&usg = AOvVaw2O82ABwOJU6-OOZAbsgwOq Retrieved June 6, 2021.

CHAPTER THREE

IoT-based infant monitoring device

Dilber Uzun Ozsahin[1,2,3], Basil Bartholomew Duwa[3,4],
John Bush Idoko[5], Mohamad Khir Hamdan[4], Ghazy Aljammal[4],
Khaldun Elsafdy[4] and Ilker Ozsahin[3,6]

[1]Department of Medical Diagnostic Imaging, College of Health Science, University of Sharjah, Sharjah, United Arab Emirates
[2]Research Institute for Medical and Health Sciences, University of Sharjah, Sharjah, United Arab Emirates
[3]Operational Research Center in Healthcare, Near East University, Nicosia/TRNC, Mersin 10, Turkey
[4]Department of Biomedical Engineering, Near East University, Nicosia/TRNC, Mersin 10, Turkey
[5]Applied Artificial Intelligence Research Center, Department of Computer Engineering, Near East University, Nicosia/TRNC, Mersin 10, Turkey
[6]Department of Radiology, Brain Health Imaging Institute, Weill Cornell Medicine, New York, NY, United States

Contents

Practical Design and Applications of Medical Devices.
DOI: https://doi.org/10.1016/B978-0-443-14133-1.00018-5

3.1 Introduction

Healthcare is currently becoming a delicate part of human lifestyle, because everyone needs to know his or her health condition, especially newborn babies, to avoid unintended risks. The high rate of sudden deaths of newborn babies as a result of heart failure, or sometimes high fever is traceable to nursing mothers' negligence. The child monitoring method has historically been conducted by full-time mothers. However, nowadays, mothers are less dedicated to their kids due to carrier engagements. Hence they are always busy and have little time to sit at home and look after their children. Some nursing mothers are opportune to live close to their parents/grandparents who serve as babysitters. But those without such privilege find this cumbersome. The other option is to employ childcare or maids as guardians. But this is pricey for most parents to afford [1].

Today, technology is making it simpler and easier to monitor children by offering various solutions to several everyday issues. Now we can fix most problems with our cell phones within a short period of time. Therefore technology is important in our everyday struggles since it has proven to solve complex problems. The use of technology in childcare facilitates to ease the stress of nursing mothers, especially those engaged in daily work. With such technology, parents can track the health status of their baby. A smartphone application that allows parents to communicate with various components, such as a camera, cardiac rate monitor, body temperature sensor, DC motor, and more, forms the basis of the proposed system. The Raspberry Pi microprocessor is a single-board machine that controls the contact. Communication takes place through WiFi technology, which ensures that parents use the device even though their child is grown up and a new one is born. When at work, this device guarantees parents peace of mind. It also helps parents to be informed each time an infant needs medication, whether he or she is ill or not. The device constantly monitors the health status of a child. If the clinical status of the child is irregular, the system immediately alerts the parents. A similar study was done on the construction of smart medical dispatchers using the same components as the ones used in this study [2].

3.1.1 Statement of problem

Nowadays, most nursing mothers have higher academic qualifications, leading them to working tirelessly to boost their family finances. To be successful as working class nursing mothers, they require a system/device to monitor their children's welfare to avoid unintended dangers. Also maids are costly and inefficient in terms of healthcare, so the safest option is for children to use infant care services because it helps them to monitor the health of the child even when at work. It is therefore possible for nursing mothers to be informed of adverse problems, such as the child's temperature or other health hazards. With the proposed system, nursing mothers can easily manage these and other children-related health problems.

3.1.2 Objectives

The objectives of the proposed system are as follows:
1. Develop an Internet of Things (IOT) infant care system.
2. Implement the infant care system using a mobile application.
3. Develop an algorithm to manage the medication time and alert the user.
4. Implement a system that allows the parent to remotely swing the cradle.
5. Implement a system that notifies the parents about imbalances in their child's health.
6. Set up a live streaming camera that allows the parents to monitor their infant at any time.

3.1.3 Scope

The baby healthcare device allows parents to use the Internet to communicate with multiple components, such as Raspberry Pi microprocessor, sensor types, as well as IP cameras. The Raspberry Pi serves as the brain of the device and performs activities between the software and hardware components. Some such sensors include the cardiac rhythm sensor, temperature sensor, sound sensor, and weight sensor, which control the medication and camera, allowing parents to see the child from anywhere at any moment. In a situation where the child encounters a health issue(s), the machine alerts the parent. Moreover, the parent ought to note medication times to keep the kid in good health. There is also a DC engine that helps the user to swing the cradle with the smartphone application.

3.2 Existing systems

In the past, parents depended on traditional babysitters or housemaids to take care of their infants, but not all families could afford such services because they were expensive and unsafe. Nowadays, with the software revolution, it is becoming easier and cheaper to take care of infants using software components. A lot of systems are there with different functionalities and are compatible with the demands of the parents, ready to replace traditional infant care, with the same costs or sometimes cheaper and safer because it is under the parents' control.

3.2.1 Sproutling Baby Monitor

This is one of the smart wearable devices controlled by mobile applications to keep track of the baby's health. This system was developed by a team of former Apple and Google engineers. And it is the company's first product. This smart device straps around an infant's ankle and measures heart rate and interprets mood through the mobile application. It can predict sleeping habits and heart rates based on past data. It also allows the parents to adjust the room temperature, sound, and light levels for optimal sleeping conditions. In addition, the Sproutling Baby Monitor determines the mood of the baby. This system also predicts the sleeping hours of the baby. All these measurements are made available to the parents via the application [3].

3.2.2 Owlet Baby Monitor

A clever wearable system developed by a group of Brigham Young University students is the Owlet Baby Monitor. It is a wearable sock with sensors capable of monitoring the well-being, heart rate, oxygen content, skin temperature, quality of sleep, and place of sleep of the baby. This information can be monitored over the Internet or on a smartphone. Some of the benefits of Owlet are that it reduces the rate of premature death in children and is fast and user-friendly. The system has some restrictions, such as being highly expensive, working only while your baby sleeps, etc. [4].

3.2.3 Mimo baby monitoring with Nest Cam

The Mimo smart baby monitoring system is placed on the child's clothing or on the crib cover. The smartphones are connected via Bluetooth to the parent. The system includes several functionalities, including the baby's sleep condition, breathing patterns, body positions, skin temperature, and live audio sent to the smart devices (iPhone, iPad, or Android). Mimo combines the video camera and the clever thermostat with other intelligent devices in the building. The temperature of the baby is shown in the application, and if there is activity inside the kid's crib, a Nest Cam takes the snapshots of the memory [5]. Some of the advantages of Mimo are that the camera has more features and allows the baby to be tracked, works perfectly well even when placed on the baby's clothes, etc. A major disadvantage of the system is that it is linked to Bluetooth so it does not have a long range.

Most of the features of several programs are supported by the proposed system. The proposed system also has more basic treatment features, such as drug alerts, dose counts, a smart cradle, and live video feeding. The system integrates the functions of watching the baby live with the camera and recording infants' well-being, in addition to crying and smart cradles, with a single smartphone and web device by monitoring the heart rate, temperature, and drug alert [6].

3.2.4 Mobile application (software platform)

A smartphone device is used to build the childcare system. Two major types of mobile operating systems have been used for the implementation of the proposed system, which are iOS and Android systems.

3.2.4.1 iOS platform

The iOS smartphone application was first unveiled on June 29, 2007. This operating system is powered by C/C++, etc. The software is owned by Apple for Apple Touch devices such as the iPhone, iPad, and iPod. Some of the advantages of iOS include awesome performance, good security, multilanguage support, good gaming and business experience, etc. The iOS platform has some disadvantages, which include a limited operating systems used only for Apple products since it is not an open-source operating system, high costs of obtaining iOS devices, and very poor battery performance [7].

3.2.4.2 Android platform

Android is a Linux kernel-based smartphone operating system. The first version of Android was launched in 2003, and Google introduced it in 2005. It is planned for use with programming languages such as C/C++, Java, etc. It was primarily designed for feedback on the touch screen and used for gaming equipment, cameras, computers, and other electronic appliances [2,8–10]. Some of the advantages of the Android platform are that it is an open-source operating system, can run with multiple tasks, has a better application market, which is the Google play store, supports Java applications, etc. It has some disadvantages, including little memory for storage, forced closing on large applications/games, has not good security, etc. [11].

3.2.5 Survey

The success of a project depends on the need for the application to be used. To illustrate the usefulness of the proposed system, a questionnaire consisting of eight questions was distributed to parents to get their opinions about the system. Some of the roles of several programs were integrated into the proposed system. Also, the system has a specific role for childcare that is not offered by the existing systems, which include medical alert, smart crib, and live camera feeding. In addition to crying detection and smart cribs, the system integrates the features between live-monitored infants via camera and tracks the health of infants by measuring heart rate, temperature, and medical syrup warnings, all in one smartphone and web system.

3.3 Implementation phase

In this phase, we tested a variety of sand boards for our monitoring method to select the safest and most effective board to produce optimal and reliable outcomes.

3.3.1 Boards

In almost every research study, the microprocessor or microcontroller is the brain behind the project, integrating hardware with software and interacting with hardware (sensors) and software, as shown in Fig. 3.1.

Different kinds of panels are there that can work with variations in requirements as microcontrollers; Arduino, Raspberry Pi, and BeagleBone.

3.3.1.1 Arduino uno

The Arduino is an open-source software and hardware-based electronics platform. The Arduino reads the system inputs and transforms them into outputs by sending those commands to the control unit to be displayed on the monitor. The programming language of Arduino is based on

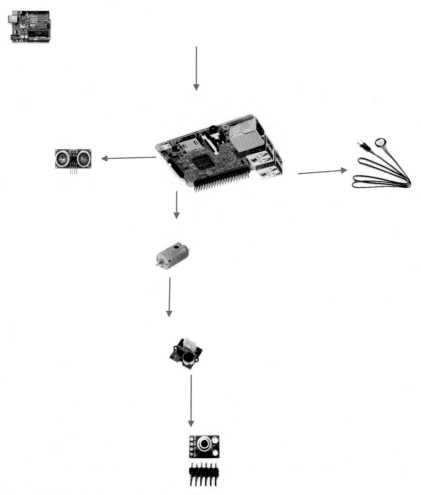

Figure 3.1 Schematic representation of the system.

wiring and Arduino (IDE) software [2]. The Arduino is ideal for a wide range of projects because it is easy and simple to work with [12].

3.3.1.2 Raspberry pi 3 model B

The Raspberry Pi is a small-size processor, like a credit card, originally created by Eben Upton for educational purposes. The concept behind Raspberry Pi's development is to create an inexpensive computer with an enhanced pre-university programming language. The Raspberry Pi is an open hardware developed to work with the Linux system. A lot of the

Linux distributions, including Raspbian, based on Debian, and Pidora, based on Fedora, have an improved version for Raspberry Pi. Some of the advantages of Raspberry Pi include low cost compared with other microprocessors, numerous add-on boards that give it more capabilities, etc. It has some disadvantages, which include the inability to perform complex multitasking, not being compatible with other operating systems like Windows, etc. [13].

3.3.1.3 Beaglebone black

This is the latest member of the BeagleBoard family, a single-board machine, like Raspberry Pi. The Sitar ARM Cortex-A8 is a low-cost, high-expansion, concentrated production package from Texas Instruments. BeagleBone Black is also supported by a wide range of other Linux distributions and operating systems, including the Bone Script library, Defiant, Apple, Ubuntu, and Cloud9 IDEs on Node.js. The BeagleBone Black boards are lightweight free hardware, and open software computers that are key products in a small box since these small PCs can be used for any form of application and perform all of the same features as a laptop or desktop. The Bone Black is made up of the BeagleBone (10/100 Ethernet port, Rest, USD Card, Micro HDMI, and DC Power). A processor, memory, and graphics acceleration are all present as chips on the board. Some of the advantages include running on multioperating systems, and having a bigger RAM compared with other microprocessors such as Raspberry Pi, etc. Its disadvantages are that it is expensive compared with other microprocessors. Moreover there are not many USB ports on the board and can only take fewer external devices at a time. Video encoding is also lacking, so you may not be able to see proper images when it is connected to a TV [14].

3.3.2 Sensors

A sensor is required to detect the amount of liquid in the bottle of medicine, which detects the level of liquid in a bottle of drink and sends alerts to consumers during medication. Heart and temperature controllers must also be used. A sound sensor is also necessary to feel the sound and warn the consumer when the baby cries. The different types of sensors are listed below, and a basic comparison determines the best sensor to be used [15].

3.3.2.1 Liquid level sensors

A water sensor brick is meant to sense water and can be used extensively in runoff, water level, and also in liquid leakage sensation. It is compact in size and can be put in small bins. Moreover, it is a user-friendly, economical, high-level/drop recognition sensor that is very sensitive. The water source for conversations based on analog output values and low power usage is also efficient and smartly built. The level of water sensor used in the container for measuring water level and the liquid touch process, such as the tape sensor, is often used in sensors that are narrower and are vertically positioned in the container in which the fluid level is measured. The drug container seems to be fine in size, but it is not helpful in medical cases because the amount of liquid is chemically sensitive and so it is sensitive to everything within the container [16].

3.3.2.1.1 Ultrasonic ranging module HC-SR04

In general, the Sonar HC-SR04 Ultrasonic Sensor calculates the distance and provides a 2cm-400cm touch measuring feature with high-precision, easy-to-use readings for sunlight or black materials. The Ultrasonic Sensor on the top of the tank determines the liquid gap between the liquid and the sensor and estimates its distance by transmitting ultrasonic signals to the fluid.

3.3.2.1.2 1 kg scale load cell weight sensor + HX711 weighing sensor

A load cell sensor is a sensor used to measure the weight of substances and has a lot of capacities. In medicine, the 1 kg capacity is used.

3.3.2.2 Sound detection sensors

3.3.2.2.1 Grove sound sensor

This sensor has a highly sensitive microphone that can sense ambient sound power. The sensor is an analog output signal, simple to use to sense vibrations, and easily combined with logic modules on the input side of the grove circuit. The L93M358 amplifier and electric microphone are analog output microphones. This sensor module helps to capture the sound (for dissemination or recording) from around your computer. The sensor is placed in infant surveillance next to the baby on the cradle to monitor the weeping waves of the baby and send corresponding alerts to the parent [17].

3.3.2.2.2 FC-04 sound sensor

This sound sensor module is a simple device to sense sound and is used for protection, switching, and surveillance applications. The precision of the condenser microphone and its ease of use can be conveniently modified for sensing. The LM393 amplifier is used for the FC-04 sensor, which is built to move rapidly from high to low, never in the center, and has an open collector power. The LM393 amplifier also provides a large 2.0 VDC to 36.0 VDC single voltage range and low performance. When the sensor senses a tone, a signal output voltage is processed. The processing is then carried out and the digitals (0 and 1; high and low) are provided to the microcontroller. The mode of use is similar to the Grove sonic sensor in baby tracking, so the sensor is positioned close to the crib and senses the child moaning sound to alert the operator when the child cries.

3.3.2.2.3 KY038-microphone module

The LM393 is used as a voltage comparator in this sensor. The same amplifier is used in the sensor FC-04 and has similar specifications. Analog and digital outputs are available. The analog output is the microphone's actual voltage output signal, and the digital output is when the sound strength is at a certain level and the output is high. The digital output is fitted with three components. First, the sensor system in front of the module measures the analog signal to the second device; and second, the amplifier improves the signal according to the resistor value of the potentiometer. A contrast between the digital output and LED when a particular value of the signal falls is the third component. By changing the potentiometer, the sensitivity can be regulated. In the same approach, the sound sensor system was used in the design of the driver's heartbeat monitor [18].

3.3.2.3 Temperature sensors
3.3.2.3.1 LM35

It is a little sensor, common and cheap, and its output voltage is linearly proportional to the temperature of the center of gravity (Celsius). The LM35 sensor is used for measuring the temperature of items. It has °C precision of 0.5°C to +25°C. It is similarly auto-heated at a low level of 0.08°C and can take the temperature between -55°C and +150°C. It has many characteristics, including sensitivity [11]. Some of the advantages include low self-heating, the capacity to measure a wide range of temperature, easy to use, and a simple circuit. Its major disadvantage is that its output is analog; hence it needs a converter to be used in boards that accept digital output to get the sensor readings.

3.3.2.3.2 MLX90614

The sensor XLX900614, often used for the sensing of temperatures, has a minimal scale and cost. This sensor has a large temperature spectrum of about 0.5°C and can be detected from -40°C up to 125°C and -70°C to 380°C for the temperature of an object. The sensor can also run at 3−5 V because of its precise functions [13]. This sensor is highly recommended to be used in healthcare projects. Some of its advantages are high accuracy, sensing temperature without contact with the object, the ability to measure moving parts, and is cheap, small, and easy to implement. Its disadvantages are that it has no liquid measurement, can be affected by humidity and dust, provides analog output, and needs a converter for further usage with other boards [8].

3.3.2.3.3 TMP36

TMP36 sensor can also work in a low voltage operation between 40°C and +125°C (2.7 V to 5.5 V). For this sensor no external calibration is required for the standard accuracy of 1°C at +25°C [14]. Based on our review of various temperature sensors, we have chosen to use the LM35 sensor. Our preference is to make the LM35 sensor smaller than others. The accuracy of the sensor is very good compared with others and is eventually cheaper than the others, as discussed above.

3.3.2.4 Heart rate sensors

3.3.2.4.1 SEN-11574

The SEN-11574 sensor is created to measure heat pulses, and is designed to operate with different microcontrollers and microprocessors. This sensor can operate with 3−5 V. The sensor has a simple design with a simple implementation procedure, and it is also easy to use. The user only needs to place it on his fingertip or earlobe to detect heartbeat. Some of the advantages of this sensor are that it is small, easy to use, and inexpensive. The major disadvantage of the sensor is that it provides only analog output and can operate only with boards that support analog output [15,16]. After studying different types of heart rate sensors, we decided to use SEN-11574, as SEN-11574, compared with other sensors, has a small and convenient size and can be placed on the infant's finger easily. Moreover, it is also cheaper compared with the heart rate sensor that provides digital output and is compatible with our Raspberry Pi [9].

3.3.2.5 DC motor

Many kinds of DC motors with various functionalities are available on the market and each motor has a specific range of power voltage to run. However, all DC motors operate together through electrical energy conversion into mechanical energy. The DC engine with 5 V power is ideal for Raspberry Pi, which provides the proposed system up to 5 V. This engine works without problems with Raspberry Pi. We chose another DC motor, which requires 9 V power to swing the cradle, but uses an external power supply as Raspberry Pi cannot supply 9 V to this motor to function [10].

3.4 Design of the proposed system

3.4.1 System overview

The proposed system helps the parents to supervise their child and contains two key elements: the hardware component and the software component (applications), as shown in Fig. 3.2. Both components communicate with each other via the Internet. The main hardware component includes Raspberry Pi 3 Model B microprocessor that interacts with different sensors to get the data (the temperature and heart rate for the infant as well as the medicine liquid level in the medicine container) from each sensor to be sent to the user. The Raspberry Pi is also responsible for

Figure 3.2 System architecture.

receiving user instructions, such as monitoring the infant, swinging the cradle, etc., and these instructions are received and forwarded to the hardware devices by Raspberry Pi. The Raspberry Pi needs an operating system and power to operate. The operating system (Raspbian) is installed on the SD memory card in addition to the power supply to provide 5 V power. The hardware also includes several instruments to assess the state of the child's health, such as a heart rate monitor, a temperature sensor, a sound sensor detector to warn the user of the weight sensor and crying of the baby in case of sickness to measure the liquid drug amount in the bottle, etc. The Raspberry Pi camera can be added to watch the child live, and a DC engine can be added to swing the cradle.

The mobile software component includes an Android operating system-based program that helps the user to connect with the system for commands or data reception. The proposed system allows parents to monitor the baby's health condition by monitoring the baby's heart rate and body temperature. If the baby's body temperature is high or if the baby weeps, the system will alert the parent. The smartphone application is also used to track the child and the crib condition. The proposed system has a special role for a sick child by keeping track of how many doses of the drug are to be given [19].

3.4.2 System data

The proposed system is an open and manageable mobile application system with internet access. The Raspberry Pi serves as the core component of the system and Stage 0 Data Flow displays the key external bodies with which it communicates. The core framework works closely to ensure the maximum reliability of the system. In addition to the IP camera, DC speed, temperature, sound and weight sensors, motor, and the administrator and the user's key components, the system data can depict the user's functionalities in the childcare environment, and the key part of the work of the framework is to connect the instructions from the user with sensors and hardware modules. The system data illustrates how user commands can be forwarded to the small processes in the system and how those processes interact with the database and the device's components. The system involves a variety of processes within the childcare system. The device receives the user's instructions and transmits them to a particular method to be analyzed and forwarded to the user's database for notification or display [20].

3.4.3 Methodology

The technique for software development is a collection of tasks and processes that identify the tech item's performance. The technique encompasses special methods and equipment at the design phase of applications. Such techniques denote the approaches for software development and assist developers in creating a successful concept, a simple schedule, and a project management phase. There are many types of model architectures that can be used, for example, waterfall, agile, spiral, scrum, prototyping, etc.. Each model has a particular procedure for different projects.

3.4.3.1 Connectivity

When the mobile application is launched, the first page is the welcome page. If the user has registered in the system, the user can directly login using the username and password. But if the user is new, the user has to create a new account in the system and insert a new username and password, after which the system checks this data for validation. If it is valid, the system will store it in the database. If it is invalid, the system will notify the user to insert the information again. After a successful login, the user is able to update the account password. After inserting the new password, the system will update it in the database. Here, the user has the ability to check the infant's health status, and the system will display the information (heart rate and temperature) to the user as a linear graph.

In this function, the user is able to manage their account — the user can add, edit, or delete an infant's profile. In the case of adding or editing, the user inserts the information and the database is updated. In case of deletion, the database will be updated upon the user's deletion confirmation. The user is also able to monitor the infant through the mobile application using the system camera. Swinging the cradle automatically is one of the functions of the proposed system, which allows the user to swing the cradle using the mobile application that connects to the DC motor controller to switch the motor on/off. The proposed system is useful in cases of illness. Here, the system provides a function that reminds the user of the medicine time by sending notifications. This function can be enabled or disabled by the user. If this function is enabled, the system requires some information about the medicine dose time and the milliliters of liquid. Once the user confirms this process, the system stores the inserted information in the database. While the system is operating, it checks the temperature, the heart rate, the sound, and the medicine liquid level in the container. When the body temperature or heart rate is abnormal, the system sends notifications to the mobile application. Hardware design is necessary to get accurate values from the sensors that provide an analog output. In the proposed system, we used Arduino Nano, which is a very small-sized Arduino that supports the analog output from the sensors because it is easier for these sensors to be implemented with Arduino and provide more accurate data. After getting the values on Arduino, we connect the Arduino to Raspberry Pi by serial cable to get these values instead of connecting the sensors directly to Raspberry Pi and converting the output from analog to digital.

3.4.3.2 Sound sensor FC-04

This sound sensor comes with three pins (OUT, GRD, and VCC). The VCC pin is connected to a 5 V power supply, GRD is connected to any ground pin, and OUT is connected to any GPIO pin on Raspberry Pi. The sound sensor is adjusted from the adjusting point with a medium range to detect a sound that is as loud as or louder than the infant's cry.

3.4.3.3 Scale load cell weight sensor with HX711

This weight sensor needs an amplifier that converts the analog data to digital to be used with HX711 amplifiers. This weight sensor has four colored wires (red, black, white, and green) to be connected to the HX711 pins (E + , E-, A-, A + ,B-, and B +) and to the other terminal of HX711; there are four pins (GND, DT, SCK, and VCC) connected to the pins on Arduino Nano: GND goes to the ground, VCC goes to the 5 V-power supply, and DT and SCK go to any of the two pins on Arduino.

3.4.3.4 Raspberry pi camera

It is easy to attach the Pi camera to Raspberry Pi. On Raspberry Pi, there are two connectors, one for the LCD screen and the other for the Pi camera. The camera is connected to Raspberry Pi by connecting the camera cable to the tiny connector labeled CSI. Subsequently, the camera module is enabled in Raspberry Pi with Python code and in some libraries for video streaming, which we are able to access from the server.

3.4.3.5 DC motor

In the proposed system, we use a 9 V DC motor; hence there is a need for an external 12 V power supply to turn on the system. We connect the

12 V power supply to a voltage regulator to reduce the power to 9 volts. After this, we connect the other terminal of the regulator to the DC motor. The Raspberry Pi is connected to the DC motor to control the process; by connecting or disconnecting the electrical current. In case of turning off Raspberry Pi or failure in the electrical power, a Flyback diode is used to keep the electrical current inside the circuit to maintain the integrity of Raspberry Pi from the opposite voltage that may lead to Raspberry Pi burning.

3.4.3.6 LM35 temperature sensor

LM35 sensor comes with three pins (VCC, GND, and Data) on its module. Instead of using the MCP converter, we use the Arduino Nano, which supports the analog output, to transfer these values to Raspberry Pi. The three pins are connected to different pins on the Arduino Nano board.

3.4.3.7 SEN-11574 heart rate sensor

SEN–11574 heart rate sensor comes with three pins (VCC, GND, and Data) on its module. The three pins are connected to different pins of the Arduino Nano board for the readings of the sensor to be transferred to the Raspberry Pi board.

3.4.4 Coding

This section discusses the actual implementation of the proposed system (mobile applications) and demonstrates how the hardware components are connected to the mobile applications to achieve the goals of the system.

3.4.4.1 Connecting to the server

The connection between the mobile application (front-end) and the server side is based on: sending data in JSON format through the API to verify, insert, update, or request data again. This data is managed on the server. The server sends a response in JSON format to the application. This is demonstrated in Algorithm 1.

Algorithm 1. Post Data to LogIn

```
 1: function login
 2: userData = user Input(username , password)
 3: if username != Empty && password != Empty then
 4:     post userData to login function in server
 5:     get response from server          •
 6:     if response == valid information then
 7:         go to home page and login successful
 8:     else
 9:         login failed invalid username or password
10:     end if
11: else
12:     Please Fill In The Required Info
13: end if
```

3.4.4.2 Server side (raspberry pi)

Raspberry Pi is used as a server for the proposed system. The PHP programming language is used to code the backend. This slim library implemented in PHP code links the application to the SQL database by receiving the JSON file sent from the application and passes this JSON information through the database to verify, insert, delete, or update the database. After that, the PHP code fetches the data as JSON and sends it back to the mobile application as a response through the API. This is depicted in Algorithm 2.

Algorithm 2. Check login data

```
 1: function login
 2: get userData from the application
 3: check userData with database   •
 4: if userData.username == valid && userData.password == valid then
 5:     send response = valid information
 6: else
 7:     send response = invalid information
 8: end if
```

3.4.4.3 Sending notifications (google cloud messaging)

The Google Cloud Messaging system is a mobile notification service developed by Google. It enables third-party application developers to send notifications from the server side to the client side. It is an Android mobile application, and once the Android device is registered on this service, the developer can send the notification using the Cloud Messaging console. In the proposed system, we implemented PHP code on the server side that communicates with the Cloud Messaging website and allows the server to send the notification automatically by passing the message to the Cloud Messaging system to be directly forwarded to the user, as demonstrated in Algorithm 3.

Algorithm 3. Send notifications through google cloud messaging

---

```
1: <?php
2: define('API_ACCESS_KEY', 'API generated by Google');      //API access key
   from Google API's Console
3: $registrationIds = array("Registerd devices' tokens generated by longing in");
4: $msg = array('message','title','Icon');      //Message Info
5: $headers = array('Authorization: key=' . API_ACCESS_KEY, 'Content-Type: ap-
   plication/json');      //API google access Authorization
6: $URL = 'https://gcm-http.googleapis.com/gcm/send'      //Google Messaging
   console URL
```

3.4.4.4 Sound sensor

The FC-04 sound sensor is easy to implement with Raspberry Pi directly without any analog to digital conversion. It operates to detect the sound frequency within a specific range (crying frequency). The Python code in Raspberry Pi, which functions to notify the user when sound is detected, is demonstrated in Algorithm 4.

Algorithm 4. Crying detection

1: frequency = get the crying frequency
2: turnSensor = off default until the user turns it on
3: **while** turnSensor == on **do**
4:      **if** frequency >= 1000 HZ **then**
5:          count++
6:          **if** count >= 10 **then**
7:              send notification
8:          **end if**
9:      **end if**
10: **end while**

3.5 Conclusion

In this research, an IoT-based monitoring device is designed and implemented. The system provides the user with the ability to track the infant's health, monitor the infant, swing the cradle, and manage the medicine doses and time using an Android mobile application. The explored sensors are the heart rate sensor to measure heartbeats, the temperature sensor to measure body temperature, the Raspberry Pi camera, the DC motor to swing the cradle, and finally the weight sensor to track the medicine liquid level and medicine time. This study explored many existing infant care systems and compared them with the proposed system. All the existing systems studied are similar in some functionalities, but the proposed system has specific functions in medical care. The average rating of the proposed system by our first-hand professionals is within the range of 85%−94%. Our future consideration is to develop the application to work on the iOS platform to increase its portability. Furthermore, the oxygen level sensor would be added to the system, and an algorithm based on machine learning would be implemented to predict the sleeping hours of the baby.

References

[1] S.W. Admin, Ky-038 microphone sound sensor module. 2017, Retrieved from http://sensorkit.en.joy-it.net/index.php?title = KY-038.

[2] D.U. Ozsahin, J.B. Idoko, M.S. Mpofu, I.R. Swalehe, I. Ozsahin, Development of medical dispatcher: a robot that delivers medicine, Modern Practical Healthcare Issues in Biomedical Instrumentation, Academic Press, 2022, pp. 97−103.

[3] T. Agarwal, Android operating system and advantages, 2023. Retrieved from http://www.edgefxkits.com/blog/android-operating-system-advantages/.

[4] B. Black, Beaglebone black, 2017. Retrieved from https://beagleboard.org/black.

[5] K. Cesternino, P.H. Strom, Analog liquid level sensor, 2001. (US Patent 6,269,695).

[6] Chang, M. Knox, T. Li, Pros and cons of raspberry pi, 2015. Retrieved from https://sites.google.com/site/mis237groupprojectraspberrypi/home/what-is-rasp-berry-pi/pros-and-cons-of-the-raspberry-pi.

[7] Device and low voltage temperature sensors, 2019. Retrieved from http://www.analog.com/media/en/technical-documentation/data-sheets/TMP353637.pdf.

[8] Sarwar, Advantages and disadvantages of using arduino, 2016. Retrieved from http://engineerexperiences.com/advantages-and-disadvatages.html.

[9] Singh, iOS vs android, 2013. Retrieved from https://www.slideshare.net/sumitvik-ram/ios-vs-android-16421573.

[10] Switzer, Density of water vapor pressure of water. Chemistry Department North Carolina State University. Retrieved from https://chemistry.sciences.ncsu.edu/resource/H2Odensityvp.html.

[11] Hazell, How to convert mg to mL (milligrams to milliliters), 2016. Retrieved from http://www.thecalculatorsite.com/articles/units/convert-mg-to-mL.php.

[12] A. Health, How to treat your child's fever. *Allina Health's Patient Education Department experts*. 2010. Retrieved from https://www.allinahealth.org/.

[13] A.M. Helmenstine, Density definition, 2017. Retrieved from https://www.thought-co.com/definition-of-density-604425.

[14] Hirshkowitz, M., Whiton, K., Albert, S.M., Alessi, C., Bruni, O., DonCarlos, L., et al. National Sleep Foundation's sleep time duration recommendations: methodology and results summary. Sleep Health. 2015;1(1):40−43.

[15] L.C. Sensor, Load cell sensor. Retrieved from http://www.loadstarsensors.com/what-is-a-load-cell.html.

[16] G. Senthilkumar, K. Gopalakrishnan, V.S. Kumar, Embedded image capturing system using raspberry pi system, International Journal of Emerging Trends a Technology in Computer Science 3 (2) (2014) 213−215.

[17] D. Seifert, Sprouting's baby monitor of the future aims to put parents at ease, 2014.

[18] D.U. Ozsahin, J.B. Idoko, B.B. Duwa, M. Zeidan, I. Ozsahin, Construction of vehicle shutdown system to monitor driver's heartbeats, Modern Practical Healthcare Issues in Biomedical Instrumentation, Academic Press, 2022, pp. 123−138.

[19] C. Technologies, Hc-sr04 user's manual, 2013. Retrieved from https://docs.google.com/document/d/1Y-yZnNhMYy7rwhAgyLpfa39RsB-x2qR4vP8saG73rE/edit#.

[20] D.U. Ozsahin, J.B. Idoko, N.T. Muriritirwa, S. Moro, I. Ozsahin, Application and impact of phototherapy on infants, Modern Practical Healthcare Issues in Biomedical Instrumentation, Academic Press, 2022, pp. 151−165.

CHAPTER FOUR

Construction of a miniaturized Covid-19 medical kit

Dilber Uzun Ozsahin[1,2,3], Declan Ikechukwu Emegano[3,4],
Omar Haider[4], Ismail Ibrahim[4], Basil Bartholomew Duwa[3,4],
Fadel Alayouti[4] and Ilker Ozsahin[3,5]

[1]Department of Medical Diagnostic Imaging, College of Health Science, University of Sharjah, Sharjah, United Arab Emirates
[2]Research Institute for Medical and Health Sciences, University of Sharjah, Sharjah, United Arab Emirates
[3]Operational Research Center in Healthcare, Near East University, Nicosia/TRNC, Mersin 10, Turkey
[4]Department of Biomedical Engineering, Near East University, Nicosia/TRNC, Mersin 10, Turkey
[5]Department of Radiology, Brain Health Imaging Institute, Weill Cornell Medicine, New York, NY, United States

Contents

4.1 Introduction

According to the World Health Organization (WHO), emergency medical kits are standardized containers or groups of items (medicines

Practical Design and Applications of Medical Devices.
DOI: https://doi.org/10.1016/B978-0-443-14133-1.00012-4

or medical supplies) put together that are needed in emergencies, crises, or disasters. These kits are designed to offer dependable medical supplies or equipment. Additionally, they are utilized by medical professionals, nongovernmental organizations, and other humanitarian organizations to get emergency access to crucial drugs and supplies [1]. These kits are useful in responding to Covid-19 patients who suffer from severe acute respiratory syndromes. Severe acute respiratory syndrome coronavirus 2 (SARS-CoV-2, formerly 2019-nCoV) is a single-stranded genomic (RNA) virus with an envelope and a positive sense [2]. Taxonomically, SARS-Cov-2 belong to the Coronavirus and Arbovirus subgenus family, both of which have several other species that can cause mild, moderate, and severe diseases in people [3]. Statistics have shown that between 2019 December, and 2020 April, 95% of gene sequences from gene banks, the National Microbiology Data Center and the National Genomics Data Center, were used in a study to characterize the whole genome of SARS-CoV-2. In this process, they utilized 30 kilobytes of genes from envelope protein (E), surface glycoprotein (S), nucleocapsid (N) proteins, and membrane protein (M) [4]. In November 2019, Wuhan, in the Hubei province of the People's Republic of China, first reported the outbreak of Covid-19. The biological transmission between humans started earlier, and most of the cases were perhaps connected with seafood [5]. The first reported case of the virus by WHO was on December 19, 2019. WHO declared the Covid-19 outbreak a pandemic because the number of infected persons was increasing tremendously across the globe [6]. As of July 15, 2020, there were about 14 million active SARS-Cov-2 cases around the world, of which 582,126 deaths occurred, while 7,881,023 patients recovered [7]. During this outbreak, the rate of infection was very high, and contact tracing became very difficult as a result of social activities such as nightclubs [8].

This study aims to design an emergency monitoring device for Covid-19 patients, which could be extremely helpful in healthcare facilities to ease the influx of emergencies. Under the guidance of WHO, several production companies manufactured kits and medical devices to meet the demands of curbing the pandemic. During Covid-19 these kits were set up in key places so that they can be used quickly when needed [9]. Front-line hospital workers were trained to use the WHO emergency health kit [10].

4.2 Related studies

The study by Motta et al. focuses on the patient-centered home-based unit and hospital-based unit. The use of a home-based device, such as a mobile phone app, enable a patient's clinical records to be transferred to the health facility. Real-time tracking of the pulse rate, temperature, oxygen saturation (SpO2), and peak expiratory flow rate is provided via a physical device created in the matrix laboratory (MATLAB®) environment located inside the health center. The major shortcoming of this is its singular use and poses a problem to the aged who cannot operate the mobile app [11]. Irawati et al. conducted a study [12] on designing a device capable of regulating oxygen volume during expiration by the patients. The device is suitable for self-isolated patients and outpatients. The result of their study showed that adequate control is made on the system automatically through the oxygen regulator, which has an error of 0.5^{-1}L per min as delta [12]. The work of Irawati does not cover other vital needs of Covid-19, it only deals with SpO2 measurement. However, this study is novel and unique in that the device measures temperature and heart rate, in addition to measuring SpO2.

4.2.1 Oxygen saturation measurement

Oxygen saturation is the measure of the quantity of hemoglobin bound to oxygen. This is done with the aid of an oximeter. This device is noninvasive when placed on a person's finger for the measurement of SpO2 [13]. This procedure is the standardized method for SpO2 in medicine and is regarded as the 5th vital sign. Health workers have to understand the need for this device, its uses, and its limitations. A perfect oxygen balance in the blood is a prerequisite for life. Low levels of arterial blood oxygen can damage organ function, such as the heart and lungs, and should be handled as soon as possible. According to the American Heart Association, persistently low oxygen levels may lead to respiratory or cardiovascular collapse. To increase blood oxygen levels, oxygen treatment can be provided [14]. During oxygenation, the blood in the lungs, for example, gets oxidized when oxygen molecules flow from the air into the bloodstream [15]. Oxygen is highly controlled in the body because hypoxia can cause many short-term problems in different organ systems. The oximeter utilizes wavelength to determine the amount of oxygenated blood compared with deoxygenated hemoglobin [16,17] based on the Beer–Lambert law of light absorption that quantifies how much light gets through a certain vascular system, such as the fingertip or the earlobe. A detector can measure the red and infrared light pulses that the probe sends out via an automated system [16].

4.2.2 Heart rate measurement

Electrocardiogram (ECG) plays a vital role in the measurement of the electrical activities of the heart [18]. ECG measures the heart rate and rhythms, recognizes emotions, identifies people based on their biometrics, and provides an accurate diagnosis of heart problems. ECG is associated with noise and artifacts during its measurement. In the preprocessing phase, the primary objective is to get rid of this noise and find the fiducial (P, Q, R, S, and T waves) points [19]. ST and QT are cardiac pathologies, especially QT, which occur when the heart muscle comparatively takes a longer time during contraction and relaxation [20]. The point is identified when comparing signals from multiple patients [21]. The sources of the noise could be electrodes, which have a low frequency of 0.15 to 0.3 Hz, and noise emanating from other medical equipment with a frequency between 100 kHz and 1 MHz [22,23]. ECG has been shown to detect Bragada disease and other hereditary arrhythmic syndromes in people who faint [24]. Left ventricular hypertrophy is the most common form of inherited cardiomyopathy. Patients who have preexisting structural cardiac illnesses are much more likely to develop malignant arrhythmias [25,26]. Ventricular arrhythmia is also a sign of a heart anomaly, especially in Covid-19 patients [27]. ECG is very important for patient care, especially in the dosage of drugs. ECG monitoring is suggested to be done in patients with electrolyte imbalances as well when they are taking QT prolonged medications simultaneously [25,28].

4.3 Components of Covid-19 emergency medical kit

The engineering component of the medical kit's contents comprises the Arduino Uno, display module, breadboard, and jumper wires. The project uses Arduino-based models to calculate the SpO_2 and heart rates and monitors the ECG of the patient. In this project, a maximum of

30,100 pulse measuring devices have been used. The ECG sensor device is used to monitor the heart rate, and for this project, the singular heart rate device AD8232 is used. The sensors for this project are for heart rate and pulse oximetry.

4.3.1 Arduino Uno

This is a microcontroller board, based mainly on the so-called ATmega328. It has 14 slots for digital input and output. Arduino Uno also contains a 16 MHz wave detector, in-circuit serial programming header, a universal serial connection (USB), a power socket, six analog inputs, and a reset button. Arduino uses a chip from Future Technology Devices International Limited, USB driver, and is also specified by the ATmega16U2 programmer, who converts the USB to serial. The Arduino board is unique because it contains devices that are designed around an Atmel AVR 8-bit or Atmel ARM microcontroller [29]. The designs currently consist of USB as an interface and 6D pins for analog input, and 14 slots for digital input and output, allowing the user to attach various extension boards. The Arduino Uno could be operated by a USB device or by an external power source [30,31]. The power that does not come from a USB port could come from a rechargeable battery or an AC-to-DC adapter. Usually, a 2.1 mm center-positive connector is plugged into the power jack on the board to connect the adapter. The ground (GND) and VIN pin headers of the power cable can be used to connect the leads from a battery. The board can run on 6 to 20 V from an external power supply. If less than 7 V is given, however, the 5 V pin might give less than 5 V and the board might be unstable [32]. A power source of more than 12 V makes the regulator too hot and causes damage to the board. The best range is between 7 and 12 V [33].

4.3.2 Pulse sensor oximeter

This is a unified sensor for pulse as well as heart measurement. It uses two light-emitting diodes (LED) of different wavelengths, a photosensitive detector emitting red and infrared radiation [34]. The colors are seen when the data are read at the tip of the patient's finger, toe, or earlobe [34]. The device is sensitive, with a highly efficient photovoltaic device (optics) and an analog mild signal that processes and detects the pulse as well as heart rate impulses [35]. The sensor is compatible with Arduino and makes it easy to record the patient's vital parameters. It is also very cost-effective. A pulse oximeter measures the absorbance of blood through the detector. When the clamp is placed on the finger, toe, or ear-lobe of the patient. A parallel beam of light that measures the oxygen level is passed from the blood to the toe or earlobe. The resultant effect on the light absorption of the oxygenated and deoxygenated blood gives the measurement. The LED in the sensors is of great significance. The infrared light is specifically for pulse measurement, while the duo measures oxygen capacity whenever the heart beats. During oxygenation, blood is pumped, and during de-oxygenation, the quantity of blood is reduced. The interval between this increase and decrease is the pulse rate, which is directly proportional to the intensity of the light emitted by infrared lights [36]. The combination of the pulse oximeter and Arduino has a simplified connection and diagram. The four pins from the sensor are connected to the power supply, while the remaining pins are connected to the trigger, GND to GND, as seen in Fig. 4.1. The oximeter is programmed with Arduino in the integrated development environment. This is displayed serially on the monitor and then copied and uploaded to the Arduino board, as shown in Fig. 4.1.

The four pins on the ultrasonic sensor are voltage common collector (VCC), trig, ECHO, and GND [37]. The VCC serves as the power supply for connecting the five pins onto Arduino. The trig ignites the impulses for the sound. There is a pause immediately at the impulses for a reflected interface, i.e., the echo, whereas the GND pin is connected to

Figure 4.1 Arduino.

the Arduino. Therefore, the trig is severely connected to the output plug, like the echo. This sensor is cheap and produces a reasonable amount of data for interpretation [38].

4.3.3 Servo motor

This is typically different from normal direct current (DC) motors because they can be positioned at an angle, especially upon receiving impulses. This sensor is fitted to the server moto at an angle of about 180 degrees. The three wires are connected individually, one to the servo motor, and the earthing, and they remain in the Arduino, as seen from the prototype [39].

4.3.4 ThingSpeak

This is another vital package used in the construction of a miniaturized Covid-19 medical kit. It is an Internet of Things (IoT) device-based platform for the aggregation, visualization, and streaming of live data in the cloud. With ThingSpeak, the data can be instantly seen on the devices. This can be executed easily online. It is a device that can be used for prototyping and other IoT analytical platforms [40].

4.3.5 Breadboard

The breadboard is used in electronic circuit development and also as wires for projects that have a microcontroller board, such as the Arduino. However, it is a difficult task to commence [41]. In this project, boards are made up of strips, which are mostly separated from the main part, and the strips used as the electrical components are the best fit for this project. Most of the strip contains five slots. A breadboard is a solderless board. Depending on the aim and duration of the usage of the component, it does not require any soldering to connect to the board (see Fig. 4.2), so it can be reused [42]. The solderable breadboard is permanently made. The disadvantage of the permanent breadboard, apart from reuse, is that errors introduced during soldering cannot be reversed or amended [43].

4.3.6 Jumper wires

Jumper wires were used to make connections between the components on the board (earlier mentioned) and the Arduino header pins. We could also use the wires to connect all the circuits for this project. Jumper wires are tiny, metallic, and used to open or close a circuit, as seen in Fig. 4.3.

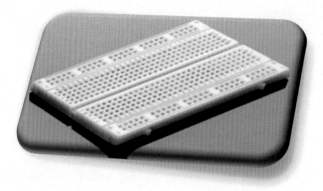

Figure 4.2 Breadboard.

Figure 4.3 Jumper wires.

The electrical circuit board is regulated by jumper wires. They come in three versions: male-to-male, male-to-female, and female-to-female. They could as well have squared or rounded heads. The male connector is the interchangeable plug, and the female is referred to as the jack [44].

4.3.7 Block and schematic diagrams

Block diagrams are highly utilized by engineers in the design of hardware, software, as well as in process flows. It is a system whereby the functions of the blocks are connected by lines that represent the relationships among the designed blocks. Schematic diagrams are usually abstract and graphical representations of the picture in reality. The major function of this

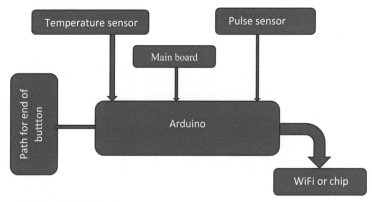

Figure 4.4 The block diagram.

diagram is the comprehensive spread of information in the abstract rather than in reality. There may be additions in the information that help the reader understand the relationships [45].

In Fig. 4.4, the block diagram shows that Arduino is the central control. The other devices are attached to it, and the whole set is viewed through a USB connection.

4.3.8 Prototype

The main Arduino + temperature sensor + pulse sensor + ESP WiFi chip is the proposed block diagram for the construction of the Covid-19 emergency kit. The temperature and pulse sensors are connected serially to the Arduino. The WiFi chip as well as the push on and off buttons are connected on either side. The first is the connection of the breadboard to the Arduino, as seen in Fig. 4.5. The temperature sensor is connected using the jumper wire via the breadboard and then to the Arduino. In the second stage, the pulse sensor is connected to the horizontal plane on the Arduino, as seen in Fig. 4.5. The WiFi chip and push-on buttons were also added to the breadboard. The final setup is made up of push son and off buttons and a Monochrome 128x32 I2C OLED graphic display.

4.3.9 Display modules

The display modules are screens that have a diameter of about 1 inch. However, because of the high contrast that organic light-emitting diode (OLED) screens have, they are easy to read, as the screen consists of 182 x 32 individual white OLED pixels, where each of these pixels is turned on

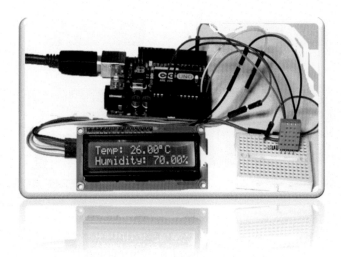

Figure 4.5 Prototype [46].

Figure 4.6 Display module prototype.

or off by a chip for the control unit. As a result, the screen provides its own lighting and does not require the presence of lighting from the back. This reduces the power needed to operate the OLED. Nevertheless, the contrast of the screen is high, which is the major focus in the choice of the screen for the project. The display module prototype is as seen in Fig. 4.6.

4.4 Conclusion

The study demonstrated the need for equipment to monitor patients in an emergency, especially during a pandemic. This equipment is designed using Arduino, a pulse oximeter, a temperature sensor, and a servo motor and incorporated into a software for effective usage. The results are displayed digitally on the galvanometer attached to the digital display with a signal chip and a graphic display. The suggested method enabled us to react swiftly to early problems in individuals. This technique may aid in preserving hospital expenditures for those in greatest need, while potentially identifying patients with severely worsening conditions. However, while the study gives a unique comparable result in the treatment of patients during Covid-19, it cannot replace the ideal equipment used in clinical settings.

References

[1] Emergency health kits. https://www.who.int/emergencies/emergency-health-kits (accessed August 18, 2022).

[2] M. Pal, G. Berhanu, C. Desalegn, V. Kandi, Severe acute respiratory syndrome coronavirus-2 (SARS-CoV-2): an update, Cureus 12 (3) (2020). Available from: https://doi.org/10.7759/CUREUS.7423.

[3] M.F. Boni, et al., Evolutionary origins of the SARS-CoV-2 sarbecovirus lineage responsible for the COVID-19 pandemic, Nature Microbiology 5 (11) (2020) 1408−1417. Available from: https://doi.org/10.1038/S41564-020-0771-4.

[4] M. Yüce, E. Filiztekin, K.G. Özkaya, COVID-19 diagnosis—a review of current methods, Biosensors & Bioelectronics 172 (2021) 112752. Available from: https://doi.org/10.1016/J.BIOS.2020.112752.

[5] K. Shen, et al., Diagnosis, treatment, and prevention of 2019 novel coronavirus infection in children: experts' consensus statement, World Journal of Pediatrics 16 (3) (2020) 223−231. Available from: https://doi.org/10.1007/S12519-020-00343-7.

[6] E.E. Isere, et al., Outcome of epidemiological investigation of COVID-19 outbreak in a south west state of Nigeria, march to august 2020, Open Journal of Epidemiology 11 (2) (2021) 163−177. Available from: https://doi.org/10.4236/OJEPI.2021.112015.

[7] COVID live—coronavirus statistics—worldometer. https://www.worldometers.info/coronavirus/ (accessed July 13, 2022).

[8] H.Y. Yuan, C. Blakemore, The impact of contact tracing and testing on controlling COVID-19 outbreak without lockdown in Hong Kong: an observational study, The Lancet Regional Health West Pacific 20 (2022). Available from: https://doi.org/10.1016/J.LANWPC.2021.100374.

[9] Emergency health kits. https://www.who.int/emergencies/emergency-health-kits (accessed July 1, 2022).

[10] H.V. Hogerzeil, J. Pinel, The new emergency health kit, Tropical Doctor 21 (1991) 47−50. Available from: https://doi.org/10.1177/00494755910210S111.

[11] L.P. Motta, et al., An emergency system for monitoring pulse oximetry, peak expiratory flow, and body temperature of patients with COVID-19 at home: development and preliminary application, PLoS One 16 (3) (2021) e0247635. Available from: https://doi.org/10.1371/JOURNAL.PONE.0247635.

[12] I.D. Irawati, S. Hadiyoso, A. Alfaruq, A. Novianti, A. Rizal, Self-oxygen regulator system for COVID-19 patients based on body weight, respiration rate, and blood saturation, Electronics (Switzerland) 11 (9) (2022). Available from: https://doi.org/10.3390/electronics11091380.

[13] (PDF) oxygen saturation. https://www.researchgate.net/publication/331407748_Oxygen_Saturation (accessed August 18, 2022).

[14] R. Beasley, et al., Thoracic society of Australia and New Zealand oxygen guidelines for acute oxygen use in adults: swimming between the flags, Respirology (Carlton, Vic.) 20 (8) (2015) 1182−1191. Available from: https://doi.org/10.1111/resp.12620.

[15] B.R. O'driscoll, R. Smith, Oxygen use in critical illness, Respiratory Care 64 (10) (2019) 1293−1307. Available from: https://doi.org/10.4187/RESPCARE.07044.

[16] L.J. Pu, Y. Shen, L. Lu, R.Y. Zhang, Q. Zhang, W.F. Shen, Increased blood glyco-hemoglobin A1c levels lead to overestimation of arterial oxygen saturation by pulse oximetry in patients with type 2 diabetes, Cardiovascular Diabetology 11 (2012). Available from: https://doi.org/10.1186/1475-2840-11-110.

[17] Oxygen saturation—StatPearls—NCBI bookshelf. https://www.ncbi.nlm.nih.gov/books/NBK525974/ (accessed July 2, 2022).

[18] R.K.C. Billones, et al., Cardiac and brain activity correlation analysis using electro-cardiogram and electroencephalogram signals, in: 2018 IEEE 10th International Conference on Humanoid, Nanotechnology, Information Technology, Communication and Control, Environment and Management, HNICEM 2018, 2019. doi: 10.1109/HNICEM.2018.8666392.

[19] ECG interpretation: characteristics of the normal ECG (P-wave, QRS complex, ST segment, T-wave) − ECG & ECHO. https://ecgwaves.com/topic/ecg-normal-p-wave-qrs-complex-st-segment-t-wave-j-point/ (accessed July 13, 2022).

[20] M. Yenikomshian, et al., Cardiac arrhythmia detection outcomes among patients monitored with the Zio patch system: a systematic literature review, Current Medical Research and Opinion 35 (10) (2019) 1659−1670. Available from: https://doi.org/10.1080/03007995.2019.1610370.

[21] M. Merone, P. Soda, M. Sansone, C. Sansone, ECG databases for biometric systems: a systematic review, Expert System with Applications 67 (2017) 189−202. Available from: https://doi.org/10.1016/J.ESWA.2016.09.030.

[22] V. Malhotra, M. Sandhu, Electrocardiogram signals denoising using improved varia-tional mode decomposition, Journal of Medical Signals and Sensors 11 (2) (2021) 100. Available from: https://doi.org/10.4103/JMSS.JMSS_17_20.

[23] A.C. Vinzio Maggio, M. Paula, E. Laciar, P. David, Quantification of ventricular repolarization dispersion using digital processing of the surface ECG, Advances in Electrocardiograms—Methods and Analysis (2012). Available from: https://doi.org/10.5772/23050.

[24] J.R. Giudicessi, R.K. Rohatgi, J.M. Bos, M.J. Ackerman, Prevalence and clinical phenotype of concomitant long QT syndrome and arrhythmogenic bileaflet mitral valve prolapse, International Journal of Cardiology 274 (2019) 175−178. Available from: https://doi.org/10.1016/J.IJCARD.2018.09.046.

[25] E.J. da S. Luz, W.R. Schwartz, G. Cámara-Chávez, D. Menotti, ECG-based heart-beat classification for arrhythmia detection: a survey, Computer Methods and Programs in Biomedicine 127 (2016) 144−164. Available from: https://doi.org/10.1016/J.CMPB.2015.12.008.

[26] M.P. Lavelle, A.D. Desai, E.Y. Wan, Arrhythmias in the COVID-19 patient, Heart Rhythm O2 3 (1) (2022) 8. Available from: https://doi.org/10.1016/J.HROO.2022.01.002.

[27] J.A. Varney, et al., COVID-19 and arrhythmia: an overview, Journal of Cardiology 79 (4) (2022) 468. Available from: https://doi.org/10.1016/J.JJCC.2021.11.019.

[28] D.R. Frisch, E.S. Frankel, D.J. Farzad, S.H. Woo, A.A. Kubey, Initial experience in monitoring QT intervals using a six-lead contactless mobile electrocardiogram in an inpatient setting, Journal of Innovations Cardiac Rhythm Management 12 (3) (2021) 4433. Available from: https://doi.org/10.19102/ICRM.2021.120301.

[29] Know arduino family—arduino project hub. https://create.arduino.cc/projecthub/Sanjay_M/know-arduino-family-f3e286 (accessed July 27, 2022).

[30] 10001073.
[31] Arduino Uno R3 development board ATmega328P with USB cable - vayuyaan. https://vayuyaan.com/shop/electronic-modules/arduino-uno-r3-development-board-atmega328p-with-usb-cable/ (accessed July 2, 2022).
[32] Arduino UNO Rev3 with long pins | arduino documentation | arduino documentation. https://docs.arduino.cc/retired/boards/arduino-uno-rev3-with-long-pins (accessed July 27, 2022).
[33] Arduino Uno.
[34] Z.J.v Cohen, S. Haxha, A. Aggoun, Pulse oximetry optical sensor using oxygen-bound haemoglobin, Optics Express 24 (9) (2016) 10115. Available from: https://doi.org/10.1364/OE.24.010115.
[35] H. Lee, et al., Toward all-day wearable health monitoring: an ultralow-power, reflective organic pulse oximetry sensing patch, Science Advances 4 (2018) 11. Available from: https://doi.org/10.1126/SCIADV.AAS9530.
[36] Using Pulse Oximeters.
[37] Ultra-Sonic 'Ping' Sensor—Arduino Project Hub. https://create.arduino.cc/projecthub/microBob/ultra-sonic-ping-sensor-a9c49e (accessed August 16, 2022).
[38] R. Electronics Corporation, Datasheet RAJ240090 / RAJ240100 3 to 10 Series Li-ion Battery Manager, 2020.
[39] Servo motor basics, working principle & theory. https://circuitdigest.com/article/servo-motor-working-and-basics (accessed August 16, 2022).
[40] IoT analytics—thingspeak internet of things. https://thingspeak.com/ (accessed August 18, 2022).
[41] Top 7 projects in robotics for beginners and intermediates - geeksforgeeks. https://www.geeksforgeeks.org/top-7-projects-in-robotics-for-beginners-and-intermediates/ (accessed July 27, 2022).
[42] M. Wolf, Interface design, Embedded System Interfacing (2019) 159—186. Available from: https://doi.org/10.1016/B978-0-12-817402-9.00008-X.
[43] Breadboard: types, connections, advantages & disadvantages. https://www.watelectronics.com/breadboard-construction-types-working/ (accessed July 2, 2022).
[44] What are jumper wires: colour, types and uses | wiltronics. https://www.wiltronics.com.au/wiltronics-knowledge-base/what-are-jumper-wires/ (accessed July 2, 2022).
[45] S. Friedenthal, A. Moore, R. Steiner, Modeling structure with blocks, A Practical Guide to SysML (2015) 115—183. Available from: https://doi.org/10.1016/B978-0-12-800202-5.00007-2.
[46] Complete guide for HC-SR04 ultrasonic sensor with arduino - circuit geeks. https://www.circuitgeeks.com/hc-sr04-ultrasonic-sensor-with-arduino/ (accessed August 28, 2022).

CHAPTER FIVE

Development of a polymerase chain reaction device

Dilber Uzun Ozsahin[1,2,3], Declan Ikechukwu Emegano[3,4], Mohammed Skaik[4], Mohammed Al Obied[4], Obada Abid[4], Basil Bartholomew Duwa[3,4] and Ilker Ozsahin[3,5]

[1]Department of Medical Diagnostic Imaging, College of Health Science, University of Sharjah, Sharjah, United Arab Emirates
[2]Research Institute for Medical and Health Sciences, University of Sharjah, Sharjah, United Arab Emirates
[3]Operational Research Center in Healthcare, Near East University, Nicosia/TRNC, Mersin 10, Turkey
[4]Department of Biomedical Engineering, Near East University, Nicosia/TRNC, Mersin 10, Turkey
[5]Department of Radiology, Brain Health Imaging Institute, Weill Cornell Medicine, New York, NY, United States

Contents

5.1 Introduction

Polymerase chain reaction (PCR) is an in vitro genetic methodology applied in the exponential enzymatic synthesis of superabundance genes, i.e., in the amplification of deoxyribonucleic acid (DNA). PCR is widely utilized in the analysis of genes. As a powerful tool, the synthesis of DNA occurs in vivo as DNA polymerase [1]. The PCR procedure was invented by scientist Mullis in 1985 as a very strong tool widely used in the analysis, changes, increases, or decreases in the expression of specific

Practical Design and Applications of Medical Devices.
DOI: https://doi.org/10.1016/B978-0-443-14133-1.00022-7

genes [2]. Isolating and analyzing individual genes is always difficult because the genes contain numerous strands in the cellular DNA [3]. The principle is centered on the replication of specific DNA that generates several billions of targeted DNA on the extracted template. These genes are amplified in large numbers to the extent that infinitesimal DNA extracts could be amplified [4]. PCR is regarded as a cloning technique; however, DNA of diverse origins is not directly analyzed because of the presence of mass nucleotides. As a result, it is first isolated and purified to get the sequences of interest, such as introns, microsatellites, and transposons, i.e., noncoded genetic materials [5]. PCR selects and then replicates these genes in their billions. The noncoding genes have no variation, but the DNA of interest will be amplified at the end of the reaction [6]. PCR detects the specialized DNA sequence in a body fluid. This allows the cloning of the acellular DNA fragments within hours through an automatic analyzer system. Recently, it is utilized in the making of fingerprints (forensic analysis) genetically, irrespective of the context [7]. Most times, it identifies varieties of animals, plants, food, etc., which are of diagnostic importance. PCR is essentially used in molecular biology. This is an essential tool used in the mutagenesis of site-directed cells. It also helps in the identification and detection of viruses and other communicable diseases, such as tuberculosis and HIV [8]. Another vital application of PCR is cloning. This technique of cloning DNA makes several copies of identical DNA, which are inserted into the genome of a larger group and then amplified [9].

5.1.1 Stages in PCR

PCR is a biological methodology that amplifies DNA in a short period of time. The amplification follows three basic steps:

1. Denaturation: A duo-stranded DNA template is heated to split the strands; denaturation is recommended at $45°C$ for $94°C-95°C$ at room temperature routinely because the DNA template contains up to 55 guanine-cytosine (GC) [10].

2. Annealing: The second stage is annealing, whereby the primers bind to the surrounding areas of the DNA. The template primers from their original to a new synthesis strand are extended from the primers [11]. The above reaction is done using deoxyribonucleic polymerase, which gives rise to specialized DNA sequences. This procedure can be repeated many times by denaturing the DNA templates to achieve the desired synthesis [12].

3. Extension: Polymerases stretch the $3'$ ends of individual primers toward the strands of the template.

The above processes are performed $25-35$ times in an exponential sequence at a hybridization temperature of $40°C-70°C$. A decrease in the temperature results in the reforming of the hydrogen bonds [13].

5.1.2 PCR device components

The PCR has a template for DNA that contains the target for amplification, the complementary primers, to the end of antisense. The PCR contains Taq polymerase with an optional temperature of $70°C$ and dNTPs. A buffered solution is provided for optimal activity of the chemical environment, which ensures the stability of the chemical reaction [14]. In addition, monovalent and bivalent cations, such as K^+ and $Mg^{2,+}$ are used in parts. Another component is bovine albumin, which binds to the PCR inhibitors. The concentrations of this bovine albumin are in the range of $10-100ug/mL$ [15]. Formaldehyde is also used to prime the template, especially in the annealing stage of the reactions. Dimethyl sulfoxide is another chemical used in PCR analysis to improve the denature of the DNA, especially in GC–rich zones [15]. The ionic strength of the mixture is usually increased by the addition of sulfates of ammonia [16].

5.1.3 Types of PCR

According to some authors, PCR is classified into PCR for single and PCR for multiple template use [17]. They can as well be classified as qualitative and quantitative PCR [18]. Qualitative PCR includes RT-PCR, nested PCR, multiplex PCR, touch-down PCR, and hot-start PCR; whereas quantitative PCR includes real-time PCR and many others. RT-PCR is used specifically in making single-stranded DNA by using transcriptase enzymes and a singular primer in the transcription process. Nested PCR uses two primers in the dual successive analysis [19]. The second primer amplifies the primer after the initial run of DNA. One advantage of this is the reduction of the usually specific binding of the products as a result of the amplification of unwanted primers. In addition, multiplex PCR [20] has several varieties of DNA that are amplified in a sequence simultaneously. However, touch-down PCR uses a higher annealing temperature at the initial stage. The polymerase synthesis commences immediately after the mixing of the reagents in a hot-stage PCR. The procedure takes effect immediately after the reaction mixture has been mixed up [21]. The disadvantage of this is the low temperature that allows nonspecific primers and products in the annealing stages. Although this can be prevented by allowing a complete temperature cycle.

Quantitative PCR, also known as real-time PCR, utilizes two major procedures to estimate the number of molecules in the assay [22]. The first phase uses standardized PCR, where the amount of analyte is viewed through gel electrophoresis against a standard. The next methodology is the quantification of the mixture during PCR processes. Examples of these are the SYBR green [23] from SYBR dyes, the quencher and reporter approach, and Taqman, which is made up of a fluorophore that is bound to an oligonucleotide (5^1) and 3^1 of the quencher.

5.1.4 Characteristics of endpoint PCR

Deoxyribonucleic acid separation is essentially the initial step throughout the PCR reaction, regardless of the category of the cell line, be it genome DNA or pathogenic bacteria plasmid DNA. For your research to go well, you need to choose nucleotide purifying products that are streamlined to get the most yield, purification, and truthfulness from almost every type of sample as well as applications [24]. Another important thing to think about is how the primer is made. Your PCR reactions must have a great design (i.e., an excellent choice of action scenes) and qualitative primers. Overall, the best length for primer pairs is between 18 and 30 nucleotides. The primers should have melting points of 65°C−75°C, and temperature between 55°C to 65°C for its efficiency [25,26]. The protein that speeds up the process is chosen. Taq polymerase is a key part of PCR because it is charged with creating new DNA strands. PCR is extremely precise and accurate, so it is essential to use superior enzymes and consumables to get the best results [27]. In addition to the aforementioned steps, thermal cycling and amplicon analysis are employed for fast, efficient, and reliable results. PCR consists of heating and cooling cycles (thermal cycling) for DNA amplification. Therefore, the success or failure of a PCR reaction may depend on selecting the right thermal cycler (which automates the process) and PCR plastics (which hold the reaction); furthermore, to the steps listed above, thermal cycling and PCR amplification analysis are used to get fast, accurate results. PCR uses cycles of cooling and heating (thermal cycling) to make more copies of DNA. So, the achievement or failure of a PCR amplification may rely heavily on choosing the correct thermal cycler (which also automatically performs the procedure) and PCR polymers (which hold the reaction) [28,29]. Nucleic acid electrophoresis is a popular step in the

PCR work process that is used to differentiate, identify, measure, and clean up amplicons. Using the appropriate equipment for nucleotide electrophoresis could indeed vastly enhance as well as speed up the results, letting you move on to applications further down the line more quickly [30,31].

5.2 Related works

The modeling of real-time (RT) PCR was based on a systematic methodology in designing the chip used in the PCR. The study points out that numerous point-of-care applications employ DNA analysis methods based on integrated circuits (ICs). Instruments for these purposes are not frequently maintained or validated. This paper proposes a systematic modeling technique for the RT-PCR machine, as device dependability is a crucial factor for regular functioning. According to the study, the suggested RT-PCR-on-chip system was predicted to perform its intended function for more than three years with comparable DNA measurement reproducibility. This demonstrates, in addition to the durability of lab-on-a-chip technology, the efficacy of the study's methodical approach to producing a robust model [32]. In his contribution to knowledge, Beutler et al. [33] designed a portable microchip used in PCR for the isolation of pathogenic organisms. The identification of pathogens at the point of care is crucial for widespread epidemic prediction for clinical issues. In this study, a fully mobile interconnected microchip-based RT-PCR machine for quick microbial identification has been constructed and designed. Three basic components, comprising heat, optics, and electronics, were merged into a mobile, small box to form the instrument. The microchip is reusable and may be replaced to accommodate varying detection requirements. Finally, we proved the excellent performance of the device by quick detection of *Salmonella* and *Escherichia coli*. The devices utilize a small sample size apart from their portability and user-friendliness [33].

5.3 Methodology

5.3.1 Experimental design

The design of this equipment is for scientific and informative purposes only, and the curiosity of learning by the students. This could not be a

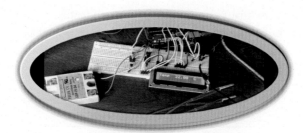

Figure 5.1 Complete setup of Arduino, breadboard, and block.

Figure 5.2 Aluminum block with holes.

substitute for any professional medical opinion. The materials needed for the design are assembled in Fig. 5.1. The methods employed in the design of this PCR originated from the design of a thermal cycler. A thermal cycler is a machine that goes from one temperature to another very quickly. To design this a holder for the test tubes is required. This holder is made up of aluminum for easy conduction of heat. This aluminum block has two holes (indicated in Fig. 5.2) for samples (depending on the number of samples) and parts for the serial connection of the wires. With two wires entering the block, the heat source from the wires flows to the resistor. It also has a cooling fan for the temperature probe and a thermo-couple that is used to connect to the thermal amplifier, which converts thermal energy to voltage energy by the Arduino control. The Arduino, the temperature display signal, and the wires are soldered to the bread-board as indicated in Fig. 5.3. The entire design is made up of material such as 12V adapter, Arduino, jumper wires and other materials as indi-cated in Fig. 5.4. The liquid cycler holds two PCR tubes and sits inside

Figure 5.3 Soldering of the wires: Arduino with the breadboard.

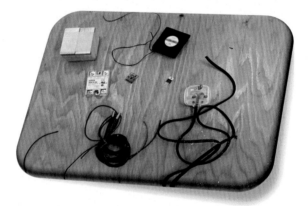

Figure 5.4 Materials for the design: Arduino, fan, resistance wires, and block.

the block. The setup is connected to an AC-to-DC power adapter. The temperature sensor (thermistor) is inserted into a hole and changes the resistance for a second. The two wires for the resistor are wired in parallel to the resistor and connected to the aluminum block. A metal block in which the samples are placed has a donor temperature of about $92°C$. Below the block are the cooling compartment fans, which cool the samples back down to an annealing temperature of $55°C-65°C$. The samples are heated again to a temperature of $72°C$ for the extension stop. This completes one PCR cycle, and the machine repeats the process $25-40$ times. The PCR is controlled by the breadboard, which is connected to the temperature sensor, heater, and cooling blocks. The aluminum conducts heat and prevents the heat from evaporating from the load or samples and condensation forming in the tubes. They can be programmed into the computer to ascertain adequate functions in the computer using the mini-PCR app. The setup is connected to the breadboard using a switch that triggers the heaters, as can be seen in Fig. 5.1. Finally, the pump is attached, which pumps water to the setup.

5.3.2 Components

This design comprises of a thermal cycler that utilizes the amplification of DNA fragments through PCR methodology. Thermal cyclers can also be employed to assist thermal reactions, such as digestion using restriction enzymes and fast diagnostic procedures. The gadget comprises a heating block with openings (holes) for inserting tubes containing reaction mixtures of samples and reagents. The temperature of the block is gradually raised and decreased in a distinct, preprogrammed cycle [34]. Prototype of mini PCR is shown in Fig. 5.5.

5.3.2.1 Fan

There are three main types of fan arrangements. They are done using two, three, or four wires. In the category of two wires, the fan turns on and off like in the three wires category, which have a sensor module. With the sensor, the frequency at which the fan spins can be evaluated. In the four-wire category, the extra wire can be used with pulse width modulation to regulate the speed of the fan. The three-wire fan incorporated with a sensor

Figure 5.5 Prototype of mini PCR [41].

is the best because the third wire is connected to the ground, mostly the yellow wire, but typically not the red or black [35].

5.3.2.2 Thermocouple

This is a wire with two dissimilar conductors. It has two metallic cables that are put together to form a section. When current is applied, the metallic strip is heated up or cooled. A current is passed during this process [36]. They measure the temperature of the sensors. The thermocouple can be a wire with two dissimilar conductors. It has two metallic cables, which are put together to form a section. Thermocouples are made up of an alloy, which is a combination of different grades of platinum and rhodium. Another type of thermocouple is the e-type made from constantan and chrome. Whereas the J-type thermocouple is formed by a combination of iron and constantan [37].

5.3.2.3 Arduino Uno

This is an ATmega328-based microcontroller board. The Arduino has 14 slots that allow digital input/output. Arduino Uno has a 16-MHz wave detector, an ICSP header, a USB connector, a power socket, six analog inputs, and a reset button. Future Technology Devices International Limited makes Arduino's USB driver chip. An ATmega16U2 programmer converts USB serial data to Arduino [38] Arduino's subcomponents are constructed around an Atmel AVR 8-bit or ARM microcontroller. USB as an interface, 6D pins for analog input, and 14 slots for digital input and output let users connect expansion boards. Arduino Uno may be powered by USB or an external source [39,40]. If a gadget has no USB port, a rechargeable battery or AC-to-DC converter may provide power. To connect an adapter, put a 2.1 mm center-positive connection into the board's power port. The connected battery leads to the power line's GND and VIN pin headers. The board may be powered by an external 6- to 20-Volt source. If less than 7 V is delivered, the 5 V pin may provide less than 5 V, causing instability [38]. Over 12 V will cause the regulator to overheat and ruin the board 7−12 Volts is ideal.

5.3.2.4 Breadboard

Breadboards, often known as plug blocks, are employed in the construction of transitory circuits. It is beneficial to engineers since it facilitates the removal and replacement of components. It is also used for the construction of circuits demonstrating the mode of operation and, afterward, the reuse of the parts in another circuit.

5.3.2.5 Temperature sensor

Generally, ICs have a temperature gauge made of a semiconductor. Two diodes of similar output power and change with temperature are employed in such detectors to track temperature changes. Those that respond in a straight line are the least accurate of the basic sensing devices. The response time of these temperature sensors is also the slowest over the smallest range of temperature ($-70°$C to $150°$C).

5.3.2.6 Prototype

The prototype for the PCR device is an advanced technology for PCR-based DNA amplification, especially the thermal cycler. In this, a tube for the reaction mixture is placed into the block's apertures for accurate temperature regulation. The fan is configured in two, three, or four wires. Using two wires, the fan's speed is regulated, three-wire fans utilize sensor modules, and four-wire fans implement pulse width modulation to create the ideal system's prototype. The prototype also uses a device called a thermocouple for monitoring system temperatures, to maintain PCR reactions that occur within the appropriate range. The thermocouple depending on the manufacturer are made from different alloys including platinum, rhodium, or constantan etc. Meanwhile, An Arduino Uno microcontroller board with 14 digital I/O pins, a 16 MHz wave detector, USB and analog inputs, and a reset button runs the PCR Thermal Cycler prototype. Additionally, the prototype's breadboard facilitates the building of quick circuits that can be dismantled and rebuilt with no effort, increasing the prototype's adaptability and providing engineers with more room to play around with different setups. On top of that, the prototype uses a temperature sensor, commonly semiconductor-based integrated circuits with diodes, to measure variations in temperatures notwithstanding moderate precision and quick responses.

5.4 Conclusion

The PCR device used by hospitals is very essential in the diagnosis of different diseases, especially viral pathogens. PCR duplicates DNA accurately and quickly. Automation in PCR boomed as a result of thermostable polymerase enzymes. Digital PCR provides quicker and more exact findings than traditional PCR; qPCR properly quantifies amplified products. PCR equipment is costly and bulky in its design. Our innovative

PCR allows more small devices to be created, especially serving as mobile laboratory, which is a user-friendly, cost-effective device that helps in the amplification and isolation of genomic DNA with fast and reliable results.

References

[1] K.B. Mullis, The unusual origin of the polymerase chain reaction, Scientific American 262 (4) (1990) 56–65. Available from: https://doi.org/10.1038/SCIENTIFICAMERICAN0490-56.

[2] M.S. Hossain, R. Ahmed, M.S. Haque, M.M. Alam, M.S. Islam, Identification and validation of reference genes for real-time quantitative RT-PCR analysis in jute, BMC Molecular Biology 20 (1) (2019). Available from: https://doi.org/10.1186/s12867-019-0130-2.

[3] R. Guo, et al., Evaluation of reference genes for RT-qPCR analysis in wild and cultivated Cannabis, Bioscience, Biotechnology, and Biochemistry 82 (11) (2018) 1902–1910. Available from: https://doi.org/10.1080/09168451.2018.1506253.

[4] K. Kadri, Polymerase chain reaction (PCR): principle and applications, Synthetic Biology - New Interdisciplinary Science (2019). Available from: https://doi.org/10.5772/INTECHOPEN.86491.

[5] E. Perenthaler, S. Yousefi, E. Niggl, T.S. Barakat, Beyond the exome: the noncoding genome and enhancers in neurodevelopmental disorders and malformations of cortical development, Frontiers in Cellular Neuroscience 13 (2019) 352. Available from: https://doi.org/10.3389/FNCEL.2019.00352/BIBTEX.

[6] B. Zhang, et al., A new dynamic deep learning noise elimination method for chip-based real-time PCR, Analytical and Bioanalytical Chemistry 414 (11) (2022) 3349–3358. Available from: https://doi.org/10.1007/s00216-022-03950-7.

[7] J. Obradovic, et al., Optimization of PCR conditions for amplification of GC-Rich EGFR promoter sequence, Journal of Clinical Laboratory Analysis 27 (6) (2013) 487. Available from: https://doi.org/10.1002/JCLA.21632.

[8] Polymerase chain reaction (PCR). https://www.ncbi.nlm.nih.gov/probe/docs/techpcr/ (accessed August 8, 2022).

[9] E. van Pelt-Verkuil, A. van Belkum, J.P. Hays, A brief comparison between in vivo DNA replication and in vitro PCR amplification, Principles and Technical Aspects of PCR Amplification (2008) 9. Available from: https://doi.org/10.1007/978-1-4020-6241-4_2.

[10] Polymerase chain reaction and its applications—microlit. https://www.microlit.us/polymerase-chain-reaction-and-its-applications/ (accessed August 10, 2022).

[11] R. Luthra, R.R. Singh, K.P. Patel (Eds.), Clinical Applications of PCR, vol. 1392, 2016. doi: 10.1007/978-1-4939-3360-0.

[12] O. Bagasra, S.P. Hauptman, H.W. Lischner, M. Sachs, R.J. Pomerantz, Detection of human immunodeficiency virus type 1 provirus in mononuclear cells by in situ polymerase chain reaction, New England Journal of Medicine 326 (21) (1992) 1385–1391. Available from: https://doi.org/10.1056/nejm199205213262103.

[13] O. Paun, P. Schönswetter, Amplified fragment length polymorphism: an invaluable fingerprinting technique for genomic, transcriptomic, and epigenetic studies, Methods in Molecular Biology 862 (2012) 75–87. Available from: https://doi.org/10.1007/978-1-61779-609-8_7.

[14] R.L. DeBiasi, K.L. Tyler, Polymerase chain reaction in the diagnosis and management of central nervous system infections, Archives of Neurology 56 (10) (1999) 1215–1219. Available from: https://doi.org/10.1001/ARCHNEUR.56.10.1215.

[15] H. Jeong, K. Oh, P. Bjorn, S. Hong, S. Cheon, Irreversible denaturation of DNA: a method to precisely control the optical and thermo-optic properties of DNA thin solid films, Photonics Research 6 (9) (2018) 918−924. Available from: https://doi.org/10.1364/PRJ.6.000918.

[16] B.C. Delidow, J.P. Lynch, J.J. Peluso, B.A. White, Polymerase chain reaction: basic protocols, Methods Molecular Biology 15 (1993) 1−30. Available from: https://doi.org/10.1385/0-89603-244-2:1.

[17] J.F.X. Wellehan, et al., Infectious disease. a retrospective review of over 200 cases, Current Therapy in Avian Medicine and Surgery (2015) 22−106. Available from: https://doi.org/10.1016/B978-1-4557-4671-2.00011-2.

[18] J.v Geisberg, Z. Moqtaderi, K. Struhl, The transcriptional elongation rate regulates alternative polyadenylation in yeast, Elife 9 (2020) 1−55. Available from: https://doi.org/10.7554/ELIFE.59810.

[19] MINT-Universal cDNA synthesis kit Cat#SK002 User Manual. Available: http://www.evrogen.com.

[20] PCR Optimization: Reaction Conditions and Components.

[21] Y. Liu, T. Wu, J. Song, X. Chen, Y. Zhang, Y. Wan, A mutant screening method by critical annealing temperature-PCR for site-directed mutagenesis, BMC Biotechnology 13 (1) (2013) 1−8. Available from: https://doi.org/10.1186/1472-6750-13-21/FIGURES/5.

[22] M.A.E. Crispim, N.A. Fraiji, S.C. Campello, N.A. Schriefer, M.M.A. Stefani, D. Kiesslich, Molecular epidemiology of hepatitis B and hepatitis delta viruses circulating in the western amazon region, North Brazil, BMC Infectious Diseases 14 (1) (2014) 1−9. Available from: https://doi.org/10.1186/1471-2334-14-94/FIGURES/2.

[23] PCR and gel electrophoresis − genetics, agriculture, and biotechnology. https://iastate.pressbooks.pub/genagbiotech/chapter/pcr-and-gel-electrophoresis/ (accessed August 10, 2022).

[24] K. Wang, T. Kadja, C. Liu, Y. Sun, V.P. Chodavarapu, Low-cost, real-time polymerase chain reaction system for point-of-care medical diagnosis, 22 (6) (2022), 2320. doi: 10.3390/22062320.

[25] DMSO: uses and risks. https://www.webmd.com/vitamins-and-supplements/dmso-uses-and-risks (accessed August 10, 2022). "Polymerase Chain Reaction."

[26] M.I. Khalil, Different types of PCR, Global Scientific Journals 9 (2) (2021) 758−768. Available from: https://doi.org/10.11216/gsj.2021.02.48315.

[27] Y. Dang, et al., Comparison of qualitative and quantitative analyses of COVID-19 clinical samples, Clinica Chimica Acta; International Journal of Clinical Chemistry 510 (2020) 613. Available from: https://doi.org/10.1016/J.CCA.2020.08.033.

[28] Y. Mo, R. Wan, Q. Zhang, Application of reverse transcription-PCR and real-time PCR in nanotoxicity research, Methods in Molecular Biology 926 (2012) 99. Available from: https://doi.org/10.1007/978-1-62703-002-1_7.

[29] M. Gaudin, C. Desnues, Hybrid capture-based next generation sequencing and its application to human infectious diseases, Frontiers in Microbiology 9 (2018). Available from: https://doi.org/10.3389/FMICB.2018.02924/FULL.

[30] M.R. Green, J. Sambrook, Touchdown polymerase chain reaction (PCR), Cold Spring Harb Protoc 2018 (5) (2018) 350−353. Available from: https://doi.org/10.1101/PDB.PROT095133.

[31] N. Marmiroli, E. Maestri, Polymerase chain reaction (PCR), Food Toxicants Analysis (2007) 147−187. Available from: https://doi.org/10.1016/B978-044452843-8/50007-9.

[32] SYBR | Thermo Fisher Scientific—TR. https://www.thermofisher.com/tr/en/home/brands/product-brand/sybr.html (accessed August 10, 2022).

[33] E. Beutler, P. Lee, Hemochromatosis, Molecular Diagnostics (2010) 177–190. Available from: https://doi.org/10.1016/B978-0-12-369428-7.00016-1.

[34] T.C. Lorenz, Polymerase chain reaction: basic protocol plus troubleshooting and optimization strategies, Journal of Visualized Experiments: JoVE 63 (63) (2012) 3998. Available from: https://doi.org/10.3791/3998.

[35] PCR Primer Design Tips—Behind the Bench. https://www.thermofisher.com/blog/behindthebench/pcr-primer-design-tips/ (accessed August 10, 2022).

[36] Biochemistry, Polymerase Chain Reaction—StatPearls—NCBI Bookshelf. https://www.ncbi.nlm.nih.gov/books/NBK535453/ (accessed August 10, 2022).

[37] J.G. Ryan, T.N. Theis, G.K. Sujan, Interconnects, Reference Module in Materials Science and Materials Engineering (2016). Available from: https://doi.org/10.1016/B978-0-12-803581-8.01803-8.

[38] Critical parameters of thermal cycling testing—Trelic | Solutions for Reliability and Materials. https://www.trelic.fi/critical-parameters-of-thermal-cycling-testing/ (accessed August 10, 2022).

[39] Y.X. Liu, et al., A practical guide to amplicon and metagenomic analysis of microbiome data, Protein and Cell 12 (5) (2021) 315–330. Available from: https://doi.org/10.1007/S13238-020-00724-8/TABLES/3.

[40] X. Dong, et al., Fast and simple analysis of MiSeq amplicon sequencing data with MetaAmp, Frontiers in Microbiology 8 (2017) 1461. Available from: https://doi.org/10.3389/FMICB.2017.01461/BIBTEX.

[41] Kuznetsov S., Matt M., Arduino PCR (thermal Cycler), https://www.instructables.com/Arduino-PCR-thermal-cycl (2023).

CHAPTER SIX

Construction of an electrophysiotherapy device

Dilber Uzun Ozsahin[1,2,3], Basil Bartholomew Duwa[3,4], Emad Shoaireb[4], Ibrahim Hashi[4], Mohamed Ismail[4] and Ilker Ozsahin[3,5]

[1]Department of Medical Diagnostic Imaging, College of Health Science, University of Sharjah, Sharjah, United Arab Emirates
[2]Research Institute for Medical and Health Sciences, University of Sharjah, Sharjah, United Arab Emirates
[3]Operational Research Center in Healthcare, Near East University, Nicosia/TRNC, Mersin 10, Turkey
[4]Department of Biomedical Engineering, Near East University, Nicosia/TRNC, Mersin 10, Turkey
[5]Department of Radiology, Brain Health Imaging Institute, Weill Cornell Medicine, New York, NY, United States

Contents

6.1 Introduction

Electrotherapy is a component of a treatment plan in which electrodes are positioned on the skin to trigger neurons, musculature, and tissues. Physical therapy is one of the most well-known ancient sciences. Humans

Practical Design and Applications of Medical Devices.
DOI: https://doi.org/10.1016/B978-0-443-14133-1.00019-7

have observed the energy-physical changes in nature as well as their significant effects on organic matter, such as the benefits and drawbacks of sunlight and UV rays. Ancient cultures, such as the Greeks, Romans, and Babylonians, utilized massage as one of the earliest known forms of physical therapy. Numerous physical treatment gadgets employing the most advanced techniques have been developed recently and continue to evolve [1].

Furthermore, one of the most common and widely used forms of electrical therapy in physical therapy is called transcutaneous electrical nerve stimulation. The transcutaneous electrical nerve stimulation unit is a piece of medical equipment that targets certain areas of the body to receive minute electrical currents from the device. The application of these currents is intended to alleviate discomfort. Because transcutaneous nerve stimulation is an efficient method that can help reduce pain and prevent pain from returning in the future, some of the units that provide this treatment are developed specifically for use in medical settings such as hospitals and other healthcare facilities [2].

However, when treating certain conditions, it is necessary to avoid using electrotherapy because it poses certain risks. The physiotherapist is the one who determines the contraindications when performing the initial evaluation of the patient. The contraindications when a person has an electrical device implanted in their body, include mental status disorder, disruption in the functioning of pacemakers or pain stimulators, malignant tissue, a shift in sensation texture, as well as excessively moist wounds [3].

6.1.1 Physiotherapy

Physiotherapy involves the utilization of heating, ice, liquid, sounds, and electric wavelengths released by the electromagnetic field, including infrared, ultraviolet, short waves, and microwaves, to induce a therapeutic response in the tissues. The primary objective of physical therapy is to rehabilitate the human body in the event of injuries, such as muscle, nerve, joint, and blood perfusion abnormalities, within a specified range of harm [4]. Physiotherapy is also categorized as:

- Mechanical and kinetic treatment: Acupuncture, cervical and spinal tension, and continuous passive movement are all forms of treatment that fall under this category.
- Thermal treatment: This comprises surface heating, deep temperature, cold, and infrared rays, in addition to water treatment.
- Treatment of the acoustics

- Phototherapy
- Electrostimulation: It is a combination of anesthesia and electrostimulation used to maintain muscle mass and avoid atrophy [5].

6.1.1.1 Electrostimulants (TENS)

Transcutaneous electrical nerve stimulation (TENS) is a type of electrical stimulation performed with a small device that sends square waves, resulting in low frequencies ranging from 0 to 200 Hz and narrow pulse amplitudes that have deeper penetration. Long pulse amplitudes, on the other hand, are used for muscle contraction and depend on high currents. This tool helps by momentarily halting the activity of nerve cells around the painful location. The body's natural painkiller, endomorphin, is also stimulated, making the TENS device useful for treating spinal injuries and acute and chronic pain just after an injury or during rehabilitation exercises [6].

6.1.1.2 Electrical muscle stimulators

These simulators are helpful during rehabilitation, especially for strengthening muscles affected by dystrophy. This is accomplished within the constraints of specific programming, which includes the following: An increase in muscular size and strength together with a reduction in subcutaneous fat have been found in athletes after long-term treatments, with the former attributed to the improvement of motor neuron output, which in turn causes adaptive changes in the contractile parts of the muscle. Electrodes can be placed in a bipolar pattern along the muscle, or in a unipolar model with one electrode on the origin of the spinal nerve root and another electrode on the motor nerve. These stimulators produce modulated or sinusoidal waves of varying pulse amplitude and frequency (50−250 Hz). Ten sessions each day, five days per week, consist of a 10−15 second stimulation period followed by a 5-second break [7].

6.2 Materials and methods

6.2.1 Working principle of the device

The device is capable of producing a wide variety of therapeutic currents that are utilized in the medical industry. This is accomplished by uploading the signals we want to generate into the memory of the controller by

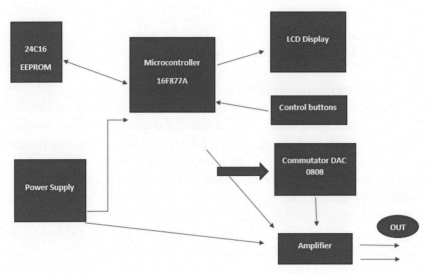

Figure 6.1 Box diagram of the device.

entering 50 consecutive values for each signal. The therapeutic currents that are produced by the apparatus can then be used in the medical field. We can select the sort of current we want to use by using the control buttons, and all of its parameters can also be changed, ranging from 0 to 500 Hz, with a pulse width of at least 250 microseconds. (Current value 1 mA up to 100 Ma, 100 V $\pm$ voltage value.)

- After delivering the start command to the micro, this signal is digitally created on the PORTD port so that the signal can be transformed into an analog signal using the DAC0808 switch so that the signal level at the output ranges to 5 V. This occurs after the start instruction has been given to the micro.
- This amplifier is a linear amplifier, which means that the signal at the output is identical to the signal at the input. The gain may be adjusted anywhere from 1 to 20. As a result, the output can swing from a minimum value of -5 V to a maximum value of $+100$ V (Figs. 6.1 and 6.2).

6.2.2 Explanation of circuit parts
6.2.2.1 Controller part
It is the main section in the circuit that controls the voter, the form, frequency, and value of the therapeutic current, as well as the display section and the activation of the integrated circuits in the device. We used the

Electro Physiotherapy Device

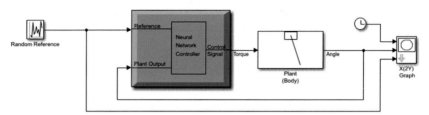

Figure 6.2 Complete circuit of the physiotherapy device.

PIC16F877A controller in our project. PIC controllers are divided into three main types.

6.2.2.1.1 PIC17CXX with high performance
These controllers are mainly used in the field of industrial technologies and depend on their work on RISC technology, which has a 16-bit width and is characterized by its ability to implement the interrupt in the shortest possible time. It has been expanded so that it contains switches, PWM modulation channels, counters, timers, and Sesame Windows.

6.2.2.1.2 Intermediate-range (PIC16FXXX)
This type provides the designer with many possibilities, as its elements have segments with legs ranging from 18 to 68, depending on the number of services available.

6.2.2.1.3 PIC18CXX
It has a 16-bit program word width and many important features, including the presence of a stack with a depth of 32 levels, in addition to multiple sources of internal and external interrupts. With the possibility of interchangeability, the digital equalizer is designed to work with a precision of 10 bits.

6.2.3 PIC controller architecture
The controller architecture is classified into two main groups:

6.2.4 CPU core control
A core is composed of the fundamental elements and operations that, together with the main processor's memory (both its program memory

and its data memory), constitute the core itself. These elements are as follows: CPU
- The memory and the arrangement of the memory
- Interrupt mechanisms
- Slide vibrator
- Keep an eye on the timer and the rest mode
- Beeping system

6.2.5 Peripheral of the controller

The surroundings of the controller are composed of two distinct sections: (1) What connects the controller with the external sign, and (2) what implements internal activities (such as timers).

The following are the different types of surroundings that are available with PIC controllers:
- O/I window for a variety of uses in general
- Timers. Timer 0, Timer 1, Timer 2
- Analog-Digital Converter (8−10 bit) ADC
- Data memory stored in an internal EEPROM device
- Communication window allowing for simultaneous serial connections
- General-purpose synchronous and asynchronous transmission systems
- Comparators, a parallel window, and an LCD screen controller

6.2.6 PIC16F87X microcontroller advantages

- It is a high-performance RISC CPU.
- It only has 25 instructions. Each of its instructions only takes one cycle of the instruction cycle, except branching instructions, which require two cycles.
- Capacity of 8 kilobytes and 14 words of flash in the program memory
- Opportunity to interrupt with 14 different techniques.
- Stacked 8 levels deep.
- Ability to either directly or indirectly solve the issue at hand
- Opportunity to reset the chip to its factory settings whenever the electrical power PRO is connected.
- A security code that can be programmed
- Existence of a timer for monitoring purposes
- Multiple choices for slide lasers
- A completely unmoving sound design

- The ability to perform serial programming on it via two programming inputs (ICSP) at a supply voltage of 5 V; this is a possibility.
- The capacity to access the memory of the program and carry out reading and writing operations
- Extensive selection of healthy 2.−5V options
- A diverse array of temperatures.
- A low overall usage of energy

6.2.7 Advantages related to controller peripherals

- Octagonal timer with divider labeled "Timer 0" (Prescaler)
- Timer (Timer 1): This timer has 16 bits and a divider that may be increased while the device is in sleep mode.
- A timer with eight bits and two dividers (Timer 2) (Prescaler, Postscaler)
- The Gateway for Synchronous Serial Communications (MSSP)
- Ten-digit digital-analog converter with an eight-bit resolution
- A synchronous and asynchronous transceiver system that is designed for general purposes.
- Sub-drive window that is 8 bits wide and has three control ports (WR, RD, CS.)
- The whistling of the therapist in the event of a severe and unexpected decline in well-nourished patients.

6.2.8 Liquid crystal display

The PIC16F877A microcontroller sends out characters, numbers, and shapes that are displayed on a liquid crystal display (LCD) screen. It has 16 inputs in total, with 6 used for controlling and 8 for transmitting data. An LCD screen measuring 16 × 2 and operating at 5 V was incorporated into our circuit (Table 6.1).

6.2.9 Power

Since the vast majority of digital integrated circuits, which include microprocessors and memories, operate with 5 V +, the power unit is considered one of the most fundamental components of any electronic system. It is an electrical transformer with an input of 220 VAC and an output of 100−12, and it delivers the supplies listed below:

- ± 5 V supply for micro, digital integrated, and commutator circuits
- A power supply with a range of − 100 V to 100 V for the drive amplifier

Table 6.1 Screen bits distribution.

Pin No	Name	I/O	Description
1	Vss	Power	GND
2	Vdd	Power	+5 V
3	Vo	Analog	Contrast control
4	RS	Input	Register select
5	R/W	Input	Read/Write
6	E	Input	Enable(Strobe)
7	DO	I/O	Data LSB
8	D1	I/O	Data
9	D2	I/O	Data
10	D3	I/O	Data
11	D4	I/O	Data
12	D5	I/O	Data
13	D6	I/O	Data
14	D7	I/O	DataMSB

6.2.10 Amplifier

It takes the signal, that is, the output by the commutator, which oscillates between +5 and −5 V, and amplifies it ten times, such that it now oscillates between ± 100 V. The regulation of the gain, and consequently the field of the output oscillation, is regulated by the variable capacitor (VC). After the integration of PWM1 pulses, the voltage that comes from the processor controls the maximum current, and the voltage that comes from the processor after PWM2 controls the minimum current.

6.3 Result

In the course of this research, a variety of different kinds of electrical currents will be created to be utilized in the treatment of different clinical illnesses employing neural networks. The digital-to-analog switch converts it into an analog signal so that it can be applied to a linear amplifier. The linear amplifier then generates the necessary current, amplitude, shape, and frequency based on discrete values that are stored in the memory of the controller. After that, the linear amplifier starts sending the analog signal at a frequency that was selected before. The circuit operation is represented in Fig. 6.3.

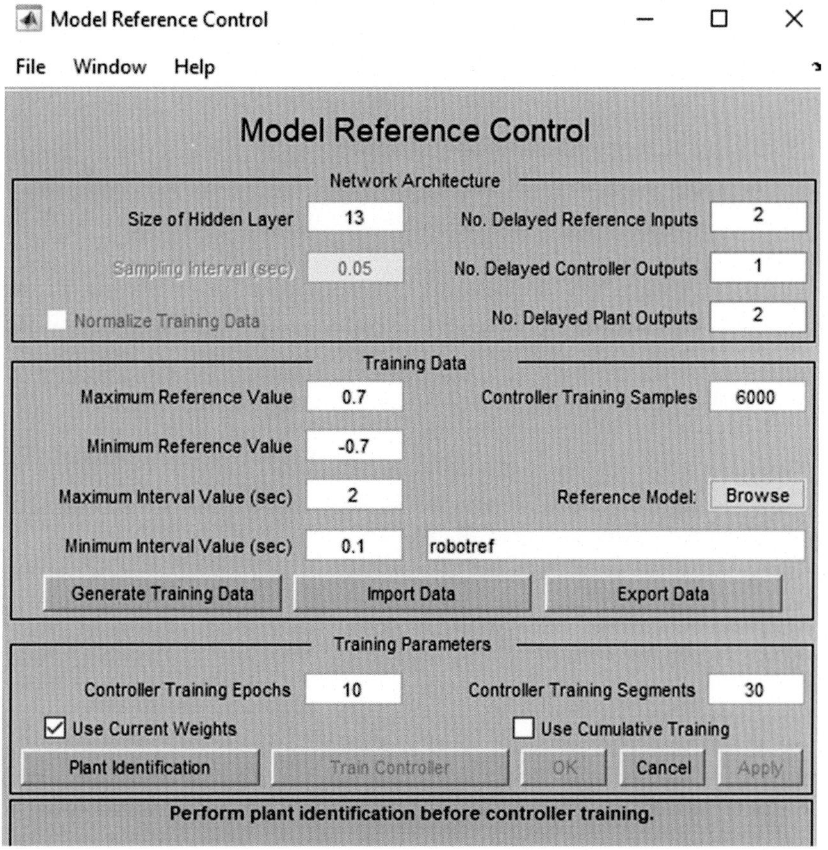

Figure 6.3 Training window in MATLAB®.

In this section, we will discuss the process of training the neural network controller to carry out its responsibilities. The first thing you need to do is import into Simulink Editor a copy of the Deep Learning Toolbox Model Reference Control block. The model reference control block must be pasted in at the second step of the process. A demonstration of the model reference controller is provided through the use of a sample model taken from the Deep Learning Toolbox software. This research focuses on the different ways in which a physiotherapy gadget can be controlled.

The next step, which would be necessary for the vast majority of situations, would be to click on Plant Identification, which would cause a new window to come up with the Plant Identification tool inside of it.

After that, the model of the plant would be given some instructions. Because the layout and operations of the Plant Identification window are the same as those of the controllers that came before it, explaining this window is superfluous.

- Click the Generate Training Data button.

The application itself is responsible for generating the data that is required for training the controller. The following window will show the point in time when the data are produced.

- Make your selection by selecting "Accept Data" from the dropdown menu. The Model Reference Control window can be used to gain access to the Train Controller. The application will train the network for a set number of repetitions using just one data segment at a time. The training set is sent out to the network one section at a time, segment by segment, until the entire set has been delivered. Training on how to operate the controller could take significantly more time than training on how to operate the plant model.

- You may return to the Model Reference Control window at this time. If the performance of the controller is inadequate, you may restart the training procedure by choosing Train Controller again. Click Generate Data or Import Data before choosing Train Controller if you want to use a new data set.

- Select Simulation > Run from the Simulink Editor's main menu to resume the simulation. This image depicts the output of a simulation and the reference signal as the simulation advances (Figs. 6.4 and 6.5).

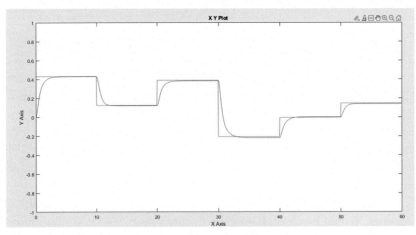

Figure 6.4 Graphical representation of the trained data.

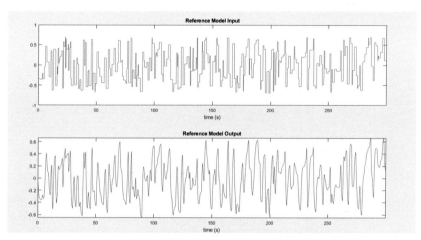

Figure 6.5 Graphical representation of the trained data showing the input and output variables.

References

[1] P. Kellaway, The part played by electric fish in the early history of bioelectricity and electrotherapy, Bulletin of the History of Medicine 20 (2) (1946) 112−137.

[2] T. Watson, The role of electrotherapy in contemporary physiotherapy practice, Manual Therapy 5 (3) (2000) 132−141.

[3] M.M.V. Miguel, I.F. Mathias-Santamaria, A. Rossato, L.F.F. Ferraz, A.M. Figueiredo-Neto, A.C. de Marco, et al., Microcurrent electrotherapy improves palatal wound healing: randomized clinical trial, Journal of Periodontology 92 (2) (2021) 244−253.

[4] K. Harman, M. Sim, J. LeBrun, J. Almost, C. Andrews, H. Davies, et al., Physiotherapy: an active, transformational, and authentic career choice, Physiotherapy Theory and Practice 37 (5) (2021) 594−607.

[5] S. Deslauriers, J. Dery, K. Proulx, M. Laliberte, F. Desmeules, D.E. Feldman, et al., Effects of waiting for outpatient physiotherapy services in persons with musculoskeletal disorders: a systematic review, Disability and Rehabilitation 43 (5) (2021) 611−620.

[6] J. Lu, Z. Liu, J. Brooks, P. Lopes, Chemical haptics: rendering haptic sensations via topical stimulants, The 34th Annual ACM Symposium on User Interface Software and Technology, Association for Computing Machinery, 2021, pp. 239−257.

[7] M.W. Kroll, P.E. Perkins, B.D. Chiles, H. Pratt, K.K. Witte, R.M. Luceri, et al., The output of electronic muscle stimulators: physical therapy and police models compared, 2021 43rd Annual International Conference of the IEEE Engineering in Medicine & Biology Society (EMBC), IEEE, 2021, pp. 1264−1268.

CHAPTER SEVEN

Voice-controlled prosthetic hand

**Dilber Uzun Ozsahin[1,2,3], Basil Bartholomew Duwa[3,4],
John Bush Idoko[5], David Edward[6], Lama Khorzom[4],
Osama Haj Hussein[4], Abdulraheem Alsiba[4], Noor Hamzah[4] and
Ilker Ozsahin[3,7]**

[1]Department of Medical Diagnostic Imaging, College of Health Science, University of Sharjah, Sharjah, United Arab Emirates
[2]Research Institute for Medical and Health Sciences, University of Sharjah, Sharjah, United Arab Emirates
[3]Operational Research Center in Healthcare, Near East University, Nicosia/TRNC, Mersin 10, Turkey
[4]Department of Biomedical Engineering, Near East University, Nicosia/TRNC, Mersin 10, Turkey
[5]Applied Artificial Intelligence Research Center, Department of Computer Engineering, Near East University, Nicosia/TRNC, Mersin 10, Turkey
[6]Department of Biochemistry, Faculty of Basic Medical Sciences, University of Jos, Jos, Nigeria
[7]Department of Radiology, Brain Health Imaging Institute, Weill Cornell Medicine, New York, NY, United States

Contents

7.1 Introduction

Currently, robotics is regarded as one of the most versatile fields in the world. In other words, this field applies to all functions of human life. It is essential for improving human interactions using incorporated models. In most cases, artificial intelligence-improved models are applied in the construction of these devices, using computerized functions to carry out

Practical Design and Applications of Medical Devices.
DOI: https://doi.org/10.1016/B978-0-443-14133-1.00024-0

human-modified activities. Various research studies have investigated the functions of artificially made limbs to enhance normal activities for disabled people. However, unique challenges are observed in these different studies based on the equipment they use [1].

An increase in the cases of accidents and injuries leads to the amputation of limbs, leaving patients unable to continue their lives in a normal way. The use of prosthetics has gained a lot of recognition around the world as it may enable patients to live almost a normal life.

The design and manufacture of artificial limbs is one of the primary tasks for specialists in the field of biomedical engineering. Its importance is reflected in its aim to reach the optimal position for the prosthesis that enables the patient to use it without complaining about an injury or discomfort, as successful rehabilitation of the amputee requires it to be practical. The acceptance of the prosthesis depends on several factors, including aesthetic aspects, weight of the prosthesis, comfort, and correct functionality [2].

From the working principle to the materials used in the manufacturing process and how the body reacts to it as it is moving and how it can be adjusted by the engineers to be used by the patient properly are some of the motivations for the proposed artificial arm. Artificial limbs are defined as the types of prostheses that replace lost limbs, such as hands or legs. The type of prosthesis used is determined by the type of amputation and the nature of the missing part of the limb [3]. Artificial limbs are required to allow patients to live a normal life with better appearances [3].

1. To remove an organ that is fully diseased or infected and may cause problems for the rest of the body.
2. The removal of a specific disease or dead tissue from the amputated organ.

7.2 Reasons for carrying out resection

7.2.1 Peripheral vascular diseases

Discontinuation of blood perfusion in the limb, whether due to illness or trauma, is an absolute indication for amputation, regardless of conditions and other factors, and cannot be maintained on one side or part of it. When its nutrition is severely affected, it will not only affect the nonuseful parts, it may even put the patient's life at risk due to the toxic products of the vandal tissues that spread throughout the body [4].

7.2.2 Injuries

Amputation may be required in the event of injuries that do not affect the circulatory system in the limb but are severe to the extent that the limb function cannot be restored, or that this function will be better when the amputation procedure happens to replace it with an artificial one [5].

7.2.3 Infections

An amputation procedure may sometimes be necessary to preserve a patient's life when the infection becomes severe, whether acute or chronic (out-of-control limb and not treatable with medication or other surgical procedures) [6].

7.2.4 Tumors

In most bone tumors, amputation is done to prevent the local recurrence of cancer, which may affect the structure of the alternative limb later [7].

7.2.5 Nerve injuries

Amputation after nerve injuries is indicated when developing trophic ulcers in the insensitive limb, and these ulcers often develop secondary infections, which aggravate the condition and compound tissue damage [8].

7.3 Development of artificially made body organs

The first recorded case of using the prosthetic device was designed for a toe for an Egyptian woman between 950–710 B.C. Using sandals was part of the old Egyptian ceremonial attire. Women especially were fancied based on their appearances in either dancing contests or attending any public gathering. Therefore, there is more importance to the design of a prosthetic toe to be used by this lady. This knowledge of the design and construction of a prosthetic toe made it possible for the subsequent development of modern artificially made body organs. With the evolution of artificial intelligence, it became more interesting to incorporate simulation techniques and independent factors into the development of self-driven devices such as prosthetics [9].

In the 1500s, artificial limbs were thought to have been invented in the form of artificial iron for a World War I veteran. Furthermore, as the war continued, there were numerous cases of amputation due to its severe nature. Many military personnel suffered injuries that required amputation. However, there was more need for redesigning the prosthetics they used, which led to finding a flexible, aesthetic, low-cost, and better device. Composite materials, plastics, and aluminum were used to form the various prosthetics they used. In the year 1946, they started using low-weight materials to make it easier for patients to reduce pressure and improve their movement. Sockets were also added to hold the prosthetic limb [10].

Similarly, in the years 1970−1990, other materials were introduced to the industry, making it a more comfortable, easy, and clean product. The materials used were light, such as plastic, polycarbonates, resin, laminates, as well as carbon fiber that was considered the lightest. By this time prosthetic limbs were costumed and fitted for each patient to provide the most pleasant and hygienic fit for each individual. Also, in the years 2000−2014, prosthetic limb designs were far more sophisticated and specialized to achieve the highest performance possible. For this purpose, responsive legs and arms for navigation, and motorized limbs controlled by sensors and microprocessors were introduced. The goal of this research is to design an artificial hand that closely mimics the natural hand in its movement, size, and dimensions. This hand is moved with voice by speaking without using any electrodes, and this movement is transmitted to the hand via motors connected to the hand by a secure transmission mechanism via cables and Arduino [11].

Nowadays, people have many choices when it comes to using prosthetics, as science interests many people with a tremendous improvement in technology in the last decade. There are similar voice-controlled devices applicable to either blind people using acoustic glasses or modernizing limbs using electrodes. Artificial hands were revealed to be developed using 3D designs to fit amputees. Therefore, limbs come in all shapes and sizes, colors, and materials as long as they meet the standards, assisting in the design and structure of the prototype. [12].

7.4 Design and implementation

This process begins with the designing and drawing of solids to be printed in layers. In other words, these drawings and shapes are converted

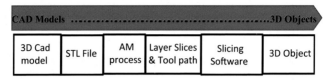

Figure 7.1 3D printing process.

into digital data, using computer program designs, such as AutoCAD, 3D Max, Rivet, etc. Then the 3D printing device produces these solids, in addition to the stabilizers, heat treatment, and further installation [13].

The first step to printing a 3D model is to build a 3D model STL file, which can be obtained from the 3D scanner, or by designing it in one of the special programs, such as 3D Max, Google Sketch, or AutoCAD. The second step is to examine the file for design errors, such as noncontacting points, as these files usually contain many of these errors, especially those taken from the 3D scanner. This process is called the correction process. In the third step, the outcome of the second step is sent as a corrected form to the Slicer Section, where this program cuts the form into a large group of very thin layers or slices that may exceed a thousand layers. The resulting file from this process is called a G–Code file, which contains a set of instructions and commands that help the printer get the job done efficiently. Afterward, the file is sent to the printer for execution based on the technology used in the printer, and at the end, the resulting stereotype enters the process of organizing and smoothing to remove the edges and unwanted parts, as demonstrated in Fig. 7.1 [14].

7.5 Materials

7.5.1 Prototype design

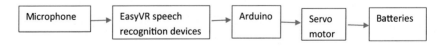

The components used in building the prosthetic arm, include Arduino, EasyVR speech recognition, microphone, servo motors, breadboard, bolts, batteries, and fishing line.

At first, we built the fingers connected by bolts; for the joints between each piece of the single finger, we used a drill to enlarge the bolts' places to make sure that they were correctly placed. Here, all the pieces were able to flex correctly with some resistance. The flexing of the fingers has the elasticity to respond to the orders (voice orders) of the patient without any error because if it does not have elasticity, it will stop working suddenly as an artifact of the artificial hand (elasticity artifact). After this, we placed the bolts and screws. We put glue on the joints we assembled except for the fingertips and made sure that the bends are all directed in the same orientation so that the finger curls. We used a fishing line because it has great strength.

We had to put the strings running above all joints in the finger and another below all the joints. The strings were inserted into the hand and out through channels, and then extended to the wrist piece. We made a double knot at the tip of the fingers so they do not slide through the two small holes. To test the elasticity of the finger, we pulled one string while holding the other. The fishing line is connected between each servo motor and the fingers. Each servo motor is responsible for the movement and strength of the finger, and we added a servo motor for the wrist movement to apply as much force as possible for a full-movement hand to the patient. The servo motor gets the electrical energy from the batteries after we successfully connected them to the electricity, and the fingers were needed to cover this stuff to have a better look with the 3D-printed covers. We did it by sticking them with glue and putting a plastic cover above the servo motors with some bolts and screws. Then the servo motors should be connected to the Arduino part and programmed correctly with one of the programming languages. After this, we continued with finger building. Here, we started work on the back side of the arm, where we separated the arm into two parts; in the first step, we had to remove/trim the appendages from all the parts using a knife, and in the second step, we assembled the two parts with glue. We used pliers to hold the parts together while gluing them. Here, we used different gluing products, such as Zap-A-Gap and epoxy. After gluing the two parts together, we re-drilled the holes to install the base that will carry the servos. Then we further trimmed the holes of the servo bed and installed it, making sure that it is completely installed in the bottom, where we had the choice of screwing the base with the hand part.

The voice input is taken into the system using the microphone connected to it. The data taken from the microphone is converted from sound waves to electrical signals so the servo motors can process it to give feedback (move to the desired location) [15]. The arm will only respond to predefined instructions since the motions of the arm are linked to the voice commands available. But new commands can be defined based on the user's needs.

Designing the hand in this way allows the user to control the hand wirelessly, which makes it very useful and easy for paralyzed people. Learning how to control it is simple, so training the user will not be an issue.

One of the main parts of this device is the microphone. In all devices some parts may start malfunctioning. In the case of the prosthetic arm, we added an application that can be downloaded on phones or tablets so that if the microphone inserted in the arm stops working, the patient will have an alternative. This application allows the patient to give voice orders for specific movements. We kept the design as simple as possible so that patients can use it easily and comfortably.

7.6 Conclusion

This research targets using voice commands to independently control a prosthetic arm. We were able to reach the desired goal, where we demonstrated the possibility and effectiveness of using sounds emanating from the microphone to direct the movement of the prosthetic arm from one point to another. Our prototype design has significantly served a purpose similar to other studies. We obtained electrical signals from sounds and conducted filtrations on this signal to constructively control arm gestures. We kept the design of the model as simple as possible so that it can be seamlessly used without complications. This product, however, could be used in industries, schools, and any public gathering.

References

[1] S. Dhanwade, R. Magar, A. Deshmukh, KC draw using arduino, International Research Journal of Engineering and Technology (IRJET) 6 (2019). Available from: https://www.irjet.net/archives/V4/i4/IRJET-V4I4443.pdf. Retrieved March 1, 2021.

[2] https://www.idosi.org/mejsr/mejsr24(TAET)16/60.pdf. Retrieved August 15, 2021.

[3] N. Jiang, S. Dosen, K.R. Muller, D. Farina, Myoelectric control of artificial limbs—is there a need to change focus?[In the spotlight], IEEE Signal Processing Magazine 29 (5) (2012) 150−152.

[4] G.C. Leng, A.J. Lee, F.G.R. Fowkers, M. Whiteman, J. Dunbar, E. Housley, et al., Incidence, natural history and cardiovascular events in symptomatic and asymptomatic peripheral arterial disease in the general population, International Journal of Epidemiology 25 (6) (1996) 1172−1181.

[5] A. Razmadze, Vascular injuries of the limbs: a fifteen-year Georgian experience, European Journal of Vascular and Endovascular Surgery 18 (3) (1999) 235—239.

[6] F. Finger, M. Rossaak, R. Umstaetter, U. Reulbach, R. Pitto, Skin infections of the limbs of Polynesian children, The New Zealand Medical Journal 117 (1192) (2004).

[7] I. Hoeber, A.J. Spillane, C. Fisher, J.M. Thomas, Accuracy of biopsy techniques for limb and limb girdle soft tissue tumors, Annals of Surgical Oncology 8 (1) (2001) 80—87.

[8] B. He, Z. Zhu, Q. Zhu, X. Zhou, C. Zheng, P. Li, et al., Factors predicting sensory and motor recovery after the repair of upper limb peripheral nerve injuries, Neural Regeneration Research 9 (6) (2014) 661.

[9] A.J. Thurston, Paré and prosthetics: the early history of artificial limbs, ANZ Journal of Surgery 77 (12) (2007) 1114—1119.

[10] P. Hernigou, AmbroiseParé IV: the early history of artificial limbs (from robotic to prostheses), International Orthopaedics 37 (6) (2013) 1195—1197.

[11] O.S. Alkhafaf, M.K. Wali, A.H. Al-Timemy, Improved Prosthetic Hand Control with Synchronous Use of Voice Recognition and Inertial Measurements, IOP Conference Series: Materials Science and Engineering 745 (2020) 012088. Available from: https://doi.org/10.1088/1757-899X/745/1/012088.

[12] D.U. Ozsahin, M. Hejazi, O.S. Adnan, H. Alloush, A. Khabbaz, J.B. Idoko, et al., Designing a 3D-printed artificial hand, Modern Practical Healthcare Issues in Biomedical Instrumentation, Academic Press, 2022, pp. 3—18.

[13] F. Ghadamli, B. Linke, Development of a desktop hybrid multipurpose grinding and 3D printing machine for educational purposes, Procedia Manufacturing 5 (2016) 1143—1153.

[14] C.K. Chua, K.F. Leong, 3D Printing and Additive Manufacturing: Principles and Applications (with Companion Media Pack) of Rapid Prototyping, *fourth ed.*, World Scientific Publishing Company, 2014.

[15] Ozsahin, D.U., Idoko, J.B., Duwa, B.B., Zeidan, M., and Ozsahin, I. (2022). Construction of vehicle shutdown system to monitor driver's heartbeats. Modern Practical Healthcare Issues in Biomedical Instrumentation 123—138 Academic Press.

CHAPTER EIGHT

3D Bioprinting of prosthetic legs

Dilber Uzun Ozsahin[1,2,3], Basil Bartholomew Duwa[3,4],
John Bush Idoko[5], Waqqas Atiq Ur Raheman Subedar[4],
David Edward[6], Janee Dinesh Barot[4] and Ilker Ozsahin[3,7]
[1]Department of Medical Diagnostic Imaging, College of Health Science, University of Sharjah, Sharjah, United Arab Emirates
[2]Research Institute for Medical and Health Sciences, University of Sharjah, Sharjah, United Arab Emirates
[3]Operational Research Center in Healthcare, Near East University, Nicosia/TRNC, Mersin 10, Turkey
[4]Department of Biomedical Engineering, Near East University, Nicosia/TRNC, Mersin 10, Turkey
[5]Applied Artificial Intelligence Research Center, Department of Computer Engineering, Near East University, Nicosia/TRNC, Mersin 10, Turkey
[6]Department of Biochemistry, Faculty of Basic Medical Sciences, University of Jos, Jos Nigeria
[7]Department of Radiology, Brain Health Imaging Institute, Weill Cornell Medicine, New York, NY, United States

Contents

8.1 Introduction

Retrospectively, limb amputation is done on individuals involved in accidents or those diagnosed with untimely diseases, such as diabetes and tumors, which require the removal of limbs in the body. Thus the construction of a prosthesis assists in substituting the limb. A myoelectric prosthesis is another form of prosthesis that uses nerve signals, involving electromyography electrodes powered by microcontrollers.

Practical Design and Applications of Medical Devices.
DOI: https://doi.org/10.1016/B978-0-443-14133-1.00020-3

When designing a prosthetic, some factors, such as cost, weight, size, power, etc., are taken into consideration. With the present innovations and introduction to 3D printing and design, the construction of prosthetics has become easier without many complications. This chapter highlights the different processes of designing prosthetics using 3D printing technology [1]. Since the turn of the century, 3D printing has proven to be a changing facet of science. However, due to the high cost involved in the existing methods, researchers are still on the lookout for an affordable yet efficient method of production of prostheses.

Usually after the amputation of a leg due to an unfortunate event, the patient often goes through rehabilitation to cope with the new state of the body. The aim of this is to allow maximum efficiency and function of the body to promote day-to-day activity. 3D printing of a prosthetic leg is not only a quick way of restoring a body part, but it also contributes greatly in boosting the confidence of a patient. The prosthesis certainly has designs that patients can customize to make it look as close as possible to the body part they had before their amputation. Moreover, 3D-printed prosthetic legs are also suitable for children, not just adults. The elite designs of several prosthetic legs allow children to express themselves in the best possible way [2].

The world of 3D printing prosthetics is new right now, and several electronics have been found. The entire procedure is in the process of taking shaping in the way prosthetics are produced by specialists. Take a look at more firm prototyping of various prosthetics with 3D-printed parts. Rapid prototyping utilizing titanium and other metals or alloys is not widespread yet, but in the coming years, it will definitely be. A period of designer prosthetics is being attempted to bring in the opportunity to manufacture affordable prosthetics. Patients with leg amputations can purchase several gadgets of varying types and switch them to match their condition [3].

8.2 Literature

The study by Biswarup and colleagues highlighted the major design techniques of a prosthetic arm that uses a gear motor control. In their block diagram, they presented the structural features of the constructed hand using a block of wood instead of an iron hand [4].Similarly, Humaid et al. reported a method of representation based on an electromyography prototype using artificial intelligence, which mimics the human hand. Omarkuluv and colleagues designed and constructed an anthropomorphic finger using a 3D-printed prototype [5].

8.2.1 Importance of prostheses

Prostheses are useful for amputees, especially for those who want to live their lives normally after an amputation due to an accident or a disease. Prostheses are beneficial for humans; therefore, this device is intended to mimic the matching character of the normal body by restoring normal functions. Similar to this study, in the design of prostheses, methods such as computer-aided design or software interface assist in analyzing the creation of a 3D graphic. Prostheses are constructed in different forms and are designed to fit into the body part that needs to be replaced.

8.2.2 3D printing of prostheses

3D printing is simply the act of making a three-dimensional object in a solidified form through a digital process. Thus the most commonly used technologies for 3D printing are selective laser sintering (SLS), fused deposition modeling (FDM), and selective stereolithography apparatus (SLA).

Based on a study by Ezigbo et al., fused filament fabrication is mostly preferred. Moreover, fused filament fabrication is cost-efficient and easy to handle compared with other devices. In fused filament fabrication printing, different materials such as plastics, organic cells, and metals, are used. The materials are acrylonitrile butadiene styrene and polylactic acid based. These are biodegradable substances with a perfume smell [6].

8.2.3 Computer-aided design

Computer-aided design (CAD) is the application of software to develop a product. CAD has different forms and different features. Some of the types of CAD available are Blender, SOLIDWORKS, Fusion 360, and AutoCAD Inventor. However, SOLIDWORKS is the preferred type due to its convenience when designing a customized printable [7].

8.3 Methods

In this study, we categorized 3D printables into three major groups: (1) control system design, (2) structural design, and (3) electronic design. The study focuses on the structured design using different analyses. In the electronic design, the electrical components of the model are used, such as the Arduino microcontroller, sensors, DC servo motors, and other fabricated components. In the design and construction of this model, we considered the replaceable characteristics of other similar devices by integrating standards into the design. The adoption of structural design and CAD tools was done for the analysis, while SOLIDWORKS was adopted in the design due to its intuitive characteristics.

SOLIDWORKS is known for its unique feature of carrying out work with convenience and ease. First, the modeling process is done by creating a 2D sketch, and then the construction is done based on a 3D sketch. Furthermore, the virtual design parts are molded, which are subsequently converted into a larger mass.

8.3.1 Design and working principles of 3D-printed prosthetic legs

In designing the feet and toes, printable components that represent the proximal phalanx and the metatarsals were constructed and linked with the thermoplastic polyurethane filaments. Similarly, artificial tendons are developed inside the structure, as shown in Fig. 8.1. The design of the leg was quite challenging because servo motors and other electronic components were all incorporated into the system. The design includes the base, lower section, and upper section. Also, spaces for screws were provided with enough materials to strongly support the objects.

8.3.2 Printing process

The fused deposition modeling method is applied after the virtual mechanical simulations. The process starts with a molten form of plastic poured into the structure-form panel to mimic the leg. Additionally, after the molten stage process, the prototype leg was transformed into the stereolithography file. The stereolithography file was then put into the software, which was subsequently converted into G-code. The G-code is important in guiding the printer based on the strength of the print head, filament extrusion velocity, and temperature.

Figure 8.1 3D prosthetic leg.

At every stage of the process, polylactic acid was loaded into the printer, which subsequently assisted the nozzle to acquire the temperature needed in the G-code and allowed the filament to reach the extrusion head; which then melted. In addition, the melted substance was now extruded in the form of a thin strand and deposited in layers, which allowed it to solidify. The process was done repeatedly in a systematic way. The whole duration of the process lasted about 24 hours.

The most common type of 3D printing involves producing items using a continuous filament of a heat-resistant plastic product. The filament is fed into a rotating, hot printer extrusion nozzle, which is collected piece by piece on the surface. The top part of the print is relocated to the subcomputer regulation to determine the structure of the produced piece. For printing on a vertical surface or paper, the top part normally works in two dimensions at a time and the top of the print is then shifted upward by a minuscule portion to launch an up-to-date surface. To halt and resume settlement, the pace of the extrusion nozzle may also be adjusted. Heat feedback from external loop heating elements influences the heat. In compliance with the temperature differential throughout the ideal number and the number recognized by the temperature sensor, the device periodically changes the energy provided to the coils, creating a negative electric circuit. The loader is pulled into this substance and warms it to the level of cooling.

Subsequently, a framework is designed to print the product first and then print the content piece by piece. To provide strength to the lingering pieces, it forms a vessel. In this step, we have to initially add the illustration to the program. It will then provide the approximate processing time. For any faults, the procedure has to be redesigned and fixed so that no mistakes are found. The processing layout will be presented based on the way the printer continues the processing of the vessel. It also creates an assist to resist and is eliminated following the completion of the printing. It takes almost 20 hours to produce the foot alone.

The 3D design program cuts the finished design into thousands and thousands of folds to create a graphical image for printing. The item can be generated in one layer at a time once the cut data is inserted into a 3D printer. Every piece (or 2D image) is interpreted by the 3D printer, and the item is formed by combining every slice with almost no visible layer indications, resulting in a 3D item.

The same equipment is not included in all 3D printing technologies. There are many methods of printing; all those accessible are composite, varying largely in the manner in which sheets are constructed to produce the output. To create the layers, some techniques use molten or weakening content. The two popular techniques using this method of printing are SLS and FDM. A further printing strategy is to print one sheet at a time when we talk about processing a photo–reactive resin utilizing a UV laser or some other equivalent energy system.

Stereolithography is the technique most commonly employed in these applications. To create the item's levels a little at a time, this method uses a container of fluid UV curable photopolymer resin and a UV beam. The light beam follows a border of the component design on the fluid resin surface of every sheet. The UV light penetration cures and strengthens the pattern formed on the resin and adds it to the coating underneath.

The elevator platform of the SLA settles at a range equal to the size of a 1 flat surface again after the design has been drawn, subsequently a wax knife brushes over the large body of the element, re-coating it with new content. The following layer pattern is mapped on this new liquid ground, overlapping the former layer. This design shapes a total of 3D items as shown in Fig. 8.2.

8.4 Conclusion

The main aim of this study was to achieve a 3D-printed prosthetic leg designed and developed using CAD software. SOLIDWORKS was

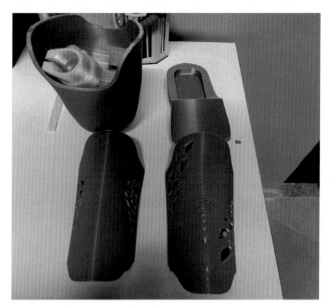

Figure 8.2 The cooling component.

adopted in the individual design of the different parts of the leg. The Xpress program was adopted in the simulation process to evaluate the performance. Additionally, the process was also adopted in the scaling and conversion of the CAD to STL. The application of 3D printing in different areas of study is not new. In this study, we focused on the simple manufacturing of prosthetics.

In our 3D-printed designed prosthetic leg, we worked tearing on the leg by installing motion sensors, which eventually led to a graph that shows the movements and how the prosthetic will move about in our future work. We would also like to show these designs to various companies that work with amputees and provide legs to those who cannot afford to buy them. The sizes will vary vastly from kids to teens and adults.

References

[1] S. Bose, S. Vahabzadeh, A. Bandyopadhyay, Bone tissue engineering using 3D printing, Materials Today 16 (2013) 496–504.
[2] J. Sun, S. Vijayavenkataraman, H. Liu, An overview of scaffold design and fabrication technology for engineered knee meniscus, Materials 10 (2017) 29.
[3] A.M. Noecker, J.F. Chen, Q. Zhou, R.D. White, M.W. Kopcak, M.J. Arruda, et al., Development of patient-specific three-dimensional pediatric cardiac models, ASAIO Journal (American Society for Artificial Internal Organs: 1992) 52 (2006) 349–353.

[4] N. Biswarup, M. Soumyajit, D. Achintya, D.N. Tibarewala, Design and implementation of the prosthetic arm using Gear Motor Control technique with appropriate testing, International Journal of Computer Application in Engineering, Technology, and Science 3 (1) (2011) 281−285.

[5] A. Humaid, J. Sayyed, L. Maozhen, Development of a local prosthetic limb using artificial intelligence, International Journal of Innovative Research in Computer and Communication Engineering 4 (9) (2016) 15708−15716. Available from: https://doi.org/10.15680/IJIRCCE.2016.

[6] P. Ezigbo, K.F. Opara, N. Chukwuchekwa, Development of 3D printable prosthetic arm for amputees using computer aided design and fused deposition modelling, International Journal of Mechatronics, Electrical and Computer Technology, 10 (2020).

[7] M.D. Fahad, M. Daniyal, S. Hassan, A. Umer, A. Emmad, A. Anees, Automation of prosthetic upper limbs for transhumeral amputees using switch-controlled motors, The International Journal of Soft Computing and Software Engineering 3 (3) (2019) 2251−7545.

CHAPTER NINE

Face recognition application in healthcare using computer web camera

Dilber Uzun Ozsahin[1,2,3], Basil Bartholomew Duwa[3,4], John Bush Idoko[5], James Gambu[6], Cemalettin Yağiz Günaşti[4], Tevfik Yavuz[4] and Ilker Ozsahin[3,7]

[1]Department of Medical Diagnostic Imaging, College of Health Science, University of Sharjah, Sharjah, United Arab Emirates
[2]Research Institute for Medical and Health Sciences, University of Sharjah, Sharjah, United Arab Emirates
[3]Operational Research Center in Healthcare, Near East University, Nicosia/TRNC, Mersin 10, Turkey
[4]Department of Biomedical Engineering, Near East University, Nicosia/TRNC, Mersin 10, Turkey
[5]Applied Artificial Intelligence Research Center, Department of Computer Engineering, Near East University, Nicosia/TRNC, Mersin 10, Turkey
[6]Department of Medical Microbiology and Clinical Microbiology, Near East University, Nicosia/TRNC, Mersin 10, Turkey
[7]Department of Radiology, Brain Health Imaging Institute, Weill Cornell Medicine, New York, NY, United States

Contents

9.1 Introduction

Digital data and image processing is a broad and promising field that is used predominantly in medicine, industry, space science, press, technology, and science and is developing day by day. With the development and widespread use of technology, computer, and telephone technologies, which are

Practical Design and Applications of Medical Devices.
DOI: https://doi.org/10.1016/B978-0-443-14133-1.00002-1
115

the blessings of technology that many of us can easily access today, keep up with the intensity and applicability of this development and research [1]. The recognition and digital processing of the human face are based on image processing techniques. Considering that technology is developing rapidly, speed, security, reliability, and ease of use are the most important elements that everyone wants to reach. Face recognition systems are used for personnel entry and exit control of large commercial companies and identification of credit card users, and are widely used in two main areas, such as the detection of legal problems, crime detection, person recognition, driver's license, and passport. If we look at the issue of identity applications in its entirety, the biggest purpose of using a face recognition system is security [2].

One of the important innovations in the development and integration of computers, technology, and security systems is image processing. The image processing technique is used to extract useful information from an image and interpret it when needed. The image to be processed can be obtained with the help of a camera or an optical scanner. To apply different processes to the image and to interpret the images in a meaningful way, different processes should be applied to the images by digitizing them. Image processing techniques are widely used in many fields, such as medicine, military, and industry. Image processing techniques are preferred in applications, such as fingerprint, iris, and face recognition for security purposes. To comment on the image taken with the aid of a scanner or camera, the images must be subjected to a processing sequence [3].

9.2 Literature review

Face recognition software is widely accepted in most industries and in medical applications, especially in the diagnosis and monitoring of diseases. İts application in various industries is estimated to be over 50 years now. Between 1964 and 1966, a team led by Woodrow Bledsoe researched on computer programming for recognizing human faces to solve problems. The team led used a scanner to map human hairline, eye and eye socket, and nose. In this study, the task of the computer was to capture this human match, but unfortunately the team was not successful. In Woodrow Bledsoe's words: "The problem of facial recognition is getting harder due to the large variation in the rotation and tilt of the head, light intensity and angle, facial expression, aging." The solution to these problems, i.e., the development of camera technology, face mapping processes, computer and machine learning, and processing speeds, have surpassed the age of face recognition. The failure of the study led scientists to new discoveries and search for new technologies [4].

With the invention and development of 2D technology, many camera systems create a flat image of a face and map points such as eye dimensions, nose, and cheekbones, which are called nodes, then the system calculates the relative position of the nodes and converts the data into a digital code. Recognition algorithms search a registered database of faces for a match. 2D camera technology systems give the best results in stable and well-lit conditions, but in dark areas, this success is severely affected. In addition, it is important that objects remain as still and stationary as possible. Unfortunately, the biggest deficit of this system is deceiving the system with photography [5]. One way to overcome the system's flaws to prevent the 2D system from being deceived by a still photograph was to detect vitality. These systems will look for indicators of a nonlive image, such as inconsistent features between the foreground and background in the image. They may ask the user to blink or move while capturing images. With the detection of vitality, criminals trying to deceive facial recognition systems by using photographs or masks can be prevented. Another revolution in facial systems is the deep convolutional neural network [6]. A convolutional neural network is applied in the detection of image patterns of the model by the user's machine or computers. The system got this name because it uses an artificial neuron network that mimics the functioning of the human brain, and in reality, this network acts like a black box. It gives the results as yet unknown input values, and then checks to make sure the network is delivering the expected result [7]. If there is a difference or problem in the result or situation, the system will make the adjustments again until it is configured properly and will be able to produce the expected outputs systematically. Today, previous advanced processes are evolving into mass-market devices. The technology is used in many fields today and one of the latest developing technologies is 3D technology. For example, 3D camera technology is used to power the thermal infrared-based face ID feature in Apple's iPhone X, which many people use today. Thermal IR imaging provides images by mapping facial models, mainly from the model of superficial blood vessels under the skin [8].

There are two types of face recognition systems: static and dynamic. While the static system creates a person recognition system from controlled and regular photographs, the dynamic system obtains image information from a video that is uncontrolled and has a fast data flow. It is possible to get the most accurate results by adjusting brightness settings, suitable filters, and threshold value settings in the systems.

While we are defining people, we can count many characteristics as personality traits. These characteristics can be given as examples of the person's age, weight, height, and eye color, and hair color, but these features can be changed quickly with safety in mind [9]. In addition, it is almost impossible for human memory to fully remember all the features it has seen, and the possibility of having more than one person with the same characteristics is another point of view. The importance of using the face recognition system becomes even more important at this point because the most prominent parts of a person's body, namely the eyes, nose, mouth, and eyebrows, cannot change without a doctor's intervention [10].

The research titled *Impaired face-like object recognition in premanifest Huntington's disease*, by Martínez-Horta et al. [11], introduces the audience to the application of face recognition objects to diagnose the presence of Huntington's disease. They applied a face-like object during an electroencephalographic process and studied on 170 components of face-like objects. The research, however, revealed increasing differences in the selected face-like objects [12−14].

9.2.1 Face recognition system algorithms

The system that identifies each person's face with a mathematical algorithm and an unchangeable and nonrepeatable code during the face recognition process, is designed to maximize the accuracy of the human face by processing the algorithm of points, such as lips and nose. The faces can be searched for matching in historical or live recordings in line with the standards determined by the institution, organization, or person using them. The source of the scanned images is not necessarily images from security cameras alone. Videos, pictures, or animated simulations on platforms, such as smartphones, cameras, or YouTube can be uploaded to the system and scanned [15]. Face recognition software examines the faces in these videos, pictures, and simulations with its advanced algorithm. If there is a blacklist prepared by the institution, organization, or person to be matched, it can detect the people on this list and give a warning. The accuracy of the solution is that it is a learning software running in the background. The system, where the camera images of the same person can be matched from different angles, can prevent erroneous selections and warnings in subsequent scans. To further increase accuracy, users can manually add contacts to the system [16]. The more images from different sources specific to the relevant person are manually uploaded to the system, the higher the learning probability and correct identification performance of the software. Ideal for places with heavy traffic such as shopping malls and stadiums, factories, and facilities with high-security levels, these systems are preparing to become an indispensable element of smart city applications that will increase in number in the near future [17].

9.2.2 Application of face recognition system

The usage areas of face recognition systems are quite wide. Today, this system, which is mostly used in airports, customs police checkpoints, and places where there is a high density of people, has the feature of being shaped according to the purpose of use. Face recognition systems can be shown as examples in schools, factories, public institutions and organizations, hospitals, shopping centers, plazas, military areas, airports, municipalities, and places with a high density of people. It is frequently used today for staff training. It is possible to trigger turnstiles, gates, and barriers with face recognition systems. [18–22].

9.3 Methodology

The research work proposes to design a Personal Disease Information Access System with a web camera using the patient's picture as the research goal. The study used the Python program and other program libraries such as OS, Numpy, Pandas, face recognition, Selenium, OpenCV, DateTime, and Matplotlib. Similarly, a website builder named YOLA was used to allow data streaming on the internet [22–24]. In addition, criteria such as age, height, weight, date of birth, infectious disease, and smoking habits were used to search on the databases for analysis. Furthermore, the following steps were followed in the research:

1. Create a specific person storage area and upload the photographs and personal information of the target audience to the storage area.
2. Determine whether they have personal identity information, Covid-19, and other health-related data stored on the system.

9.3.1 Some of the explored libraries

OS: The OS module is a module that comes ready-made in Python, allowing us to easily perform operations on files and directories. Considering that the operating systems of different operating systems are different from each other, the OS module offers many functions that enable us to communicate with different operating systems in a consistent manner.

Numpy: Numpy is a math library that allows us to do scientific calculations quickly. Also, unlike Python lists, Numpy arrays must be homogeneous, i.e., all elements in the array must be of the same data type.

Pandas: Pandas is one of the open-source Python libraries that offer easy-to-use data structures and enable data analysis. Due to its many features, it provides great convenience to those working in the field of machine learning in the data reading and preprocessing stages.

Face_Recognition: It is a library to recognize maps of the faces in the code.

Selenium: Selenium is a tool that allows you to open a browser, such as Chrome or Firefox, on the screen with the help of a driver you install on your computer and run all the processes you want with the help of a programming language like a real person.

OpenCV: It is an open-source image processing library written to create a common infrastructure for computer vision applications developed by Intel.

DateTime: It consists of some classes that offer us various functions and attributes so that we can do operations based on time and date. There are three different classes within this module.

Matplotlib: It is used to construct pictures by using numerical values.

The system designed using these libraries automatically enters the website created for this research and obtains the required information, as shown in Fig. 9.1.

From the instructions in Fig. 9.1, the def getpatient() function is defined to connect this interface to the designed website using the instruction number. Since the browser dimension is small, we increased these dimensions to the maximum level using instruction line number 18. We typed the names of the people we entered on our website in H5 font; the determining element to call those names. We wrote the contact information in P font on our website. These instructions helped in pulling the information. We learned the selector of H5 fonts with the review button on the website and saved it as people. We did the same for the P font and introduced it to our program as information. We created a space to store the information that comes in rows 24, 25, 26, and 27.

```
13
14
15    def getpatient():
16
17        driver.get("https://tevmusyavgen.yolasite.com/")
18        driver.maximize_window()
19
20
21        people = driver.find_elements_by_css_selector(".ws-section.ws-surface.ws-dark-1")
22        informations = driver.find_elements_by_css_selector(".ws-section.ws-surface.ws-light-0")
23
24        patients = []
25        patientsForDisplaying = []
26        infos = []
27        sex = []
28
```

Figure 9.1 Data information.

```
28
29        for i in people:
30
31            nameComing = i.find_element_by_css_selector('.ws-m-rich-text.ws-m-blockquote-with-icon').text
32            Tr2Eng = str.maketrans("çğıöşüÇĞİÖŞU", "cgiosuCGİOSU")
33            nameComing = nameComing.translate(Tr2Eng)
34
35            seperated = nameComing.upper().split('-')
36            justName1 = seperated[0].split()
37
38            patientsForDisplaying.append(seperated[0])
39            patients.append("".join(justName1))
40            sex.append(seperated[1])
41            # print(patients)
42
43        for j in informations:
44            infos.append(j.find_element_by_css_selector('.ws-m-rich-text.ws-m-blockquote-with-icon').text)
45
46        return patients, infos, sex, patientsForDisplaying
47
48
49
50    def texter(name,data,sex):
51        bg = cv2.imread('background.png')
52
53        ndata = data.split('/')
54
55        cv2.putText(bg,name,(40,80),cv2.FONT_HERSHEY_TRIPLEX,0.9,(0, 0, 0),2)
56        cv2.putText(bg,sex,(720,80),cv2.FONT_HERSHEY_TRIPLEX,0.9,(0, 0, 0),2)
```

Figure 9.2 Name coming.

As shown in Fig. 9.2, in the 31st row, the web addresses of the names are selected one by one and assigned to the "name coming" for every "i."

On line 32 and 33, our code converts the Turkish characters seen on our website into English characters. And thus there is no incompatibility.

In the 35th and 36th rows, name information does not come to the code by only names but more. So that we can use all different information individually, we separate that information. Afterward, the information is assigned to the lists in the 39th and 40th rows. The 43rd and 44th rows do the same thing for patient information. Finally, in the 46th row, the function returns all those valuable data so that they can be used whenever needed.

In the 50th row, the texter function is defined by using def texter(). In the texter function, we display all the infortmation coming from the database.

Fig. 9.3 shows the representation of the selenium library in the database. We used a WebDriver to reach the database. To do that we use a WebDriver, which is introduced by using the Selenium library in the 70th row.

In the 71st row, we call the getpatient function, and all the returns will be assigned to the lists on the left-hand side. As our work with the WebDriver is done, we can close it, and it can be seen in the 72nd row.

To capture a video, we call a function from the cv2 library, and also create a "while loop" to collect the information from the image sources and compare them one after the other with the live picture on the opened camera. The collected data, with the help of these lines of instructions, generated the image in Fig. 9.4.

9.4 Results

The results in Fig. 9.5 shows that by using image sources that we have imported, functions that we have defined, more functions from cv2, Face_Recognition, Numpy libraries, and finally the live camera, it becomes viable to get face detection. In addition to that, when the picture between the camera and the sources matches up, the break loop on the 152nd row starts doing its job and breaks the while loop.

```
58        for i in range(1,len(ndata)+1):
59
60            ndata[i-1] = ndata[i-1].strip()
61            belower = 80*i+80
62            cv2.putText(bg,ndata[i-1],(40,belower),cv2.FONT_HERSHEY_TRIPLEX,0.9,(0, 0, 0),2)
63
64
65        cv2.imshow("window", bg)
66        cv2.waitKey(0)
67
68
69
70     driver = webdriver.Chrome()
71     [patients,infos,sexes,namesToBeDisplayed] = getpatient()
72     driver.close()
```

Figure 9.3 Selenium library.

Figure 9.4 Face recognition.

```
165
166     sayac = 0
167
168     for k in range(0,len(patients)):
169         if (patients[k] == name):
170             print(infos[k])
171             break
172         sayac += 1
173
174
175     texter(namesToBeDisplayed[sayac],infos[sayac],sexes[sayac])
```

```
135
136             if matches[matchIndex]:
137                 name = classNames[matchIndex].upper()
138                 # print(name)
139                 y1,x1,y2,x2 = faceLoc
140                 y1,x1,y2,x2 = 4*y1, 4*x1, 4*y2, 4*x2
141                 cv2.rectangle(img,(x1,y1),(x2,y2),(0,255,0),2)
142                 cv2.rectangle(img,(x1,y2-35),(x2,y2),(0,255,0),cv2.FILLED)
143                 cv2.putText(img,name,(x1+6,y2-6),cv2.FONT_HERSHEY_COMPLEX,1,(255,255,255),2)
144                 # readInfo(name)
145                 num += 1
146
147
148         cv2.imshow('Webcam', img)
149         cv2.waitKey(1)
150
151         if num == 2:
152             break
153
```

Figure 9.5 cv 2.

The camera gets opened when the program starts, and recognizes the person looking at the camera. Now, we need to display this person's information (name, gender, age, blood type, allergenics, etc.) on the screen. Consequently, the "for loop" in the 168th row starts the searching process from the information in the database (there are lots of patients, so we need to find the right person). In the 169th row, the "if loop" asks whether we can find the information of the right person. If the answer is yes, then it breaks the for loop on the 169th row by using the variable we have defined as SAYAC. After the while loop, we have one more row to run, which is the texter function. When we run this function, by definition, it displays the person's information on the screen. Then, the code ends, as represented in Fig. 9.6.

At first we start the process of writing the 'py.\Project.py' command to the terminal for code execution. The result of the executed code is shown in Fig. 9.7–9.10, which displays the information of some of the authors of this paper.

And finally, if the image of our patient on the camera and the image in the database match, the patient information is reflected on the screen waiting for further support from healthcare professionals in emergency situations.

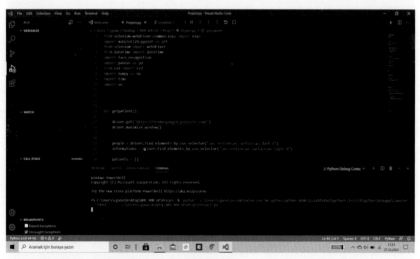

Figure 9.6 Concluding section of the coding.

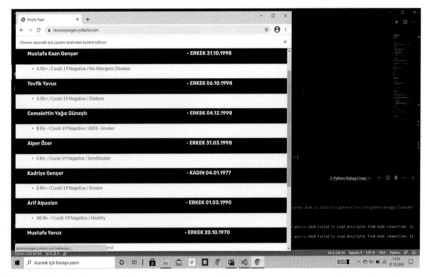

Figure 9.7 Coding results.

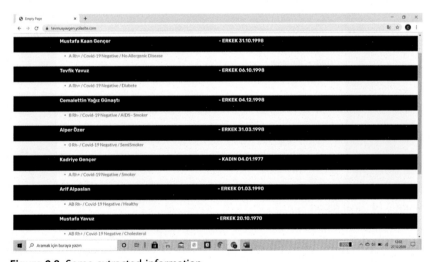

Figure 9.8 Some extracted information.

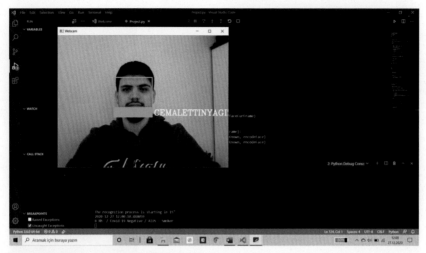

Figure 9.9 More extracted information.

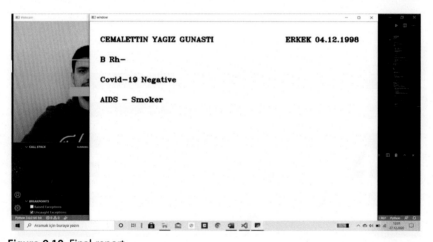

Figure 9.10 Final report.

9.5 Conclusion

This research work focuses on the modern assessment of using novel research methods to handle depressing factors in global health. It focuses on developing self-efficient computer-programmed methods to detect, identify, and get data on patients' well-being using computer programs.

Using the Python program, we chose the priority program libraries installed to call the necessary information and run smoothly. These program libraries are OS, Numpy, Pandas, Face_Recognition, Selenium, OpenCV, DateTime, and Matplotlib. We used a site builder called Yola (Open Access Program), to stream data over the internet. We archived photos of some people with distinct facial features to enable us to identify the facial features of the people we previously captured. We assigned the correct names to the photographs by entering the human information we placed in the folders completely and correctly. We checked the photos in the folder with the photos of people we took both from the internet and the camera. In the field of health systems, the infrastructure of face recognition has not been fully harnessed and this is one of the motivations for this research. In the future, bad experiences by patients can be avoided by advancing this idea and integrating it into the healthcare management systems. This could reduce the turnaround time of doctor-to-patient examination in the hospital. We hope that in the future, the integration of the face recognition system with the healthcare system will be achieved and more efficient healthcare services will be provided to patients.

References

[1] H. Wechsler, in: M.K. Kundu;, S. Mitra (Eds.), *Reliable Face Recognition Methods*: System Design, Implementation and Evaluation, Springer Science & Business Media, 2009, p. 12. ISBN 9780387384641.
[2] S. Bebbington, What is Programming. Tumblr, Archived from the original on April 29, 2020, 2014, retrieved 03.03.14.
[3] J.L. Chabert, A History of Algorithms, From the Pebble to the Microchip Springer Science & Business Media, 2012, pp. 7−8s.
[4] K. De Leeuw, J. Bergstra, The History of Information Security: A Comprehensive Handbook, Elsevier, Amsterdam, 2007, pp. 264−265.
[5] S.K. Chen, Y.H. Chang, International Conference on Artificial Intelligence and Software Engineering (AISE2014), DEStech Publication Inc, 2014, p. 21.
[6] K.K. Malay, M. Sushmita, M. Debasis, K.P. Sankar (Eds.), *Perception and machine intelligence*: First Indo-Japan Conference, PerM, in 2012, Kolkata, India, January 12−13, 2011, Springer Science & Business Media, 2012, pp. 29, ISBN 9783642273865.
[7] M.V. Valueva, N.N. Nagornov, P.A. Lyakhov, G.V. Valuev, N.I. Chervyakov, Application of the residue number system to reduce hardware costs of the convolutional neural network implementation, Mathematics and Computers in Simulation, 177, *Elsevier BV*, 2020, pp. 232−243.
[8] M.K. Bhowmik, K. Saha, S. Majumder, G. Majumder, A. Saha, A.N. Sarma, M. Nasipuri, Thermal infrared face recognition—a biometric identification technique for robust security system, Reviews, Refinements and New Ideas in Face Recognition 7 (2011) 113−138.
[9] M.A. Palmer, N. Brewer, R. Horry, Understanding gender bias in face recognition: effects of divided attention at encoding, Acta Psychologica 142 (3) (2013) 362−369.

Available from: https://doi.org/10.1016/j.actpsy.2013.01.009. ISSN 0001−6918. PMID 23422290.

[10] D.E. King, Dlib-mL: a machine learning toolkit, Journal of Machine Learning Research 10 (Jul) (2009) 1755−1758.

[11] S. Martínez-Horta, A. Horta-Barba, J. Perez-Perez, M. Antoran, J. Pagonabarraga, F. Sampedro, et al., Impaired face-like object recognition in premanifest Huntington's disease, Cortex; a Journal Devoted to the Study of the Nervous System and Behavior 123 (2019) 162−172. February 2020.

[12] R. Kimmel, G. Sapiro, The mathematics of face recognition. SIAM News, 36 (3) (2003). Archived from the original on July 15, 2007, retrieved 30.04.03.

[13] M. Haghighat, M. Abdel-Mottaleb, *Low resolution face recognition in surveillance systems using discriminant correlation analysis*, in: 2017 12th IEEE International Conference on Automatic Face & Gesture Recognition (FG 2017), 2017, pp. 912−917. doi:10.1109/FG.2017.130. ISBN 978-1-5090-4023-0. S2CID 36639614.

[14] D. Zhang, A. Jain, *Advances in Biometrics:* International Conference, ICB 2006, Hong Kong, China, January 5−7, 2006, in: Proceedings. Berlin: Springer Science & Business Media, 2006, pp. 183. ISBN 9783540311119.

[15] R. Gross, J. Shi, J. Cohn, Quo Vadis Face Recognition: Third Workshop on Empirical Evaluation Methods in Computer Vision, Carnegie Mellon University, Pittsburgh, USA, 2001. Available from: https://www.nist.gov/programs-projects/face-recognition-grand-challenge-frgc.

[16] S.E. Mason, Age and gender as factors in facial recognition and identification, Experimental Aging Research 12 (3) (2007) 151−154. Available from: https://doi.org/10.1080/03610738608259453.

[17] K.M. Jyotsna, B. Debika, *Emerging Technology in Modelling and Graphics: Proceedings of IEM Graph 2018*, Springer, 2019, pp. 672. ISBN 9789811374036.

[18] D. Sabbagh, Facial Recognition Technology Scrapped at King's Cross Site, *The Guardian*, 2019. ISSN 0261-3077, retrieved 2.09.19.

[19] P. Grother, G. Quinn, P.J. Phillips, Report on the Evaluation of 2D Still-Image Face Recognition Algorithms, National Institute of Standards and Technology, 2011, pp. 106.

[20] R. Leenes, R.V. Brakel, S. Gutwirth, P.D. Hert, *eds.* Data Protection and Privacy: The Internet of Bodies, Bloomsbury Publishing, 2018p. 176.

[21] K. Conger, R. Fausset, S.F. Kovaleski, San Francisco Bans Facial Recognition Technology, *The New York Times*, 2019. ISSN 0362−4331. *Retrieved March 26, 2020.*

[22] A. Hern, Face Masks Give Facial Recognition Software an Identity Crisis, The Guardian, 2020, ISSN 0261−3077, retrieved 24.08.20.

[23] A.J. Izenman, Linear determinant analysis, Modern Multivariate Statistical Techniques, Springer Texts in Statistics. Springer, New York. NY, 2013, pp. 237−280.

[24] A. Albiol, D. Monzo, A. Martin, J. Sastre, A. Albiol, Face recognition using HOG−EBGM, Pattern Recognition Letters 29 (10) (2008) 1537−1543.

CHAPTER TEN

Detection of retinal blood clots in the eye using laser doppler technology

Dilber Uzun Ozsahin[1,2,3], Basil Bartholomew Duwa[3,4], John Bush Idoko[5], Mohamad Sharaf Eddin[4] and Ilker Ozsahin[3,6]

[1]Department of Medical Diagnostic Imaging, College of Health Science, University of Sharjah, Sharjah, United Arab Emirates
[2]Research Institute for Medical and Health Sciences, University of Sharjah, Sharjah, United Arab Emirates
[3]Operational Research Center in Healthcare, Near East University, Nicosia/TRNC, Mersin 10, Turkey
[4]Department of Biomedical Engineering, Near East University, Nicosia/TRNC, Mersin 10, Turkey
[5]Applied Artificial Intelligence Research Center, Department of Computer Engineering, Near East University, Nicosia/TRNC, Mersin 10, Turkey
[6]Department of Radiology, Brain Health Imaging Institute, Weill Cornell Medicine, New York, NY, United States

Contents

Practical Design and Applications of Medical Devices.
DOI: https://doi.org/10.1016/B978-0-443-14133-1.00014-8
129

10.1 Introduction

Retinal or eye blood clotting can be identified as a red spot in the eye. This condition could be a result of an injury or other health complications. In medicine this condition is known as a subconjunctival hemorrhage or hyposphagma. In some individuals, a blood clot is perceived to be hemodynamic, which is an indication of many eye diseases. Thrombosis inside the blood vessels of the eye is known as disseminated intravascular coagulation; a complex disorder that results from a deficiency or problems with platelets. Vision loss occurs in many cases as a result of a temporary decrease in the perfusion of the retinal artery, the ophthalmic course, or the ciliary hall circulatory system, which prompts a deficiency in the retinal cycle, consequently causing retinal hypoxia. Retinal hypoxia is one of the most common causes of emboli. The laser Doppler technology is used to measure blood flow in the retina of the subject's eye and to detect blood clots formed as a result of any changes in the diameter of the retinal vessels. The variation in the diameters of the blood vessels occurs as part of the independent control of blood flow in healthy bodies and at different stages of the pulse cycle, while stable changes may also indicate the presence of some diseases. One of the objectives of this research is to detect blood clots formed in the retina by using Doppler technology enhanced with image processing techniques to form an accurate identification of blood clots. The Doppler scale provides important information about the blood flow in the retina so that it is placed confidently over the vessel and remains above the retina. In addition to providing continuous real-time information on blood flow, measuring the diameter of the vessel is important in regulating blood flow, as is measuring blood clots and their location to direct the laser beam in the Doppler laser system [1].

10.2 Anatomy of the eye

The structure of the human eye contains layers, the motor muscles of the eye sphere, as well as the nerves, eye vessels, orbit, and the optic pathway. The eyeball consists of two sections, front and back, and the anterior-posterior diameter is 24 mm, the cross-section is 21 mm, and it weighs 7.5 kg.

10.2.1 Cornea

A transparent front sextant, the cornea, forms the outer layer of the eye, allowing light to pass through it and refracting it on its front and back surfaces to form a real inverted imagination with the help of the eye crystal on the retina. Its refractive index is 40; with a fraction and radius convex to its front surface of about 7.7 mm.

10.2.2 Sclera

The sclera is the white, fibrous layer that forms the solid outer wall of the eye. It consists of connective tissue that is opaque to light in all areas of the back of the eyeball, but it changes somewhat in the front, becoming transparent and more convex and forming a structure known as the cornea [2].

10.2.3 Choroid

This is an analogy of the spider's brain, which is rich in pigments and in a large number of instinctive blood vessels that penetrate it, ensuring blood supply to the outer part of the retina. It consists of layers of blood vessels that vary in thickness according to their section.

10.2.4 Ciliary body

The ciliary body is a muscular epithelial vascular ring that is 6 mm wide. The ciliary processes form folds that secrete the aqueous humor that nourishes the ocular layers, such as the lens and the cornea, and provides food exchange.

10.2.5 Iris

The iris is the colored part of the eye located vertically in front of the lens and behind the cornea. It forms the front of the anterior chamber and the back of the posterior chamber of the eye. It contains an opening in its center called the pupil. It also contains smooth, circular muscle fibers with radial or longitudinal positioning.

10.2.6 Retina

The retina is an extension of the brain, reaching the ball of the eye. The sense of vision is transmitted by the optic nerve to the brain's vision centers. It is a thin transparent membrane that is surrounded by the placenta from the outside and nourishes it. The fibers of its inner layer terminate to form the optic nerve, which appears during fundoscopy in the form of a papilla or optic disk with a diameter of 1.5 mm. To the outer side of it is a depression surrounded by two vascular arcs, the macula, with a diameter of 6—5 mm, which is the center of the retina, where the fovea occupies its center and is responsible for acute vision [3].

10.2.6.1 Blood puns of the retina
The retina is the inner layer of the eye, which is a light-sensitive membrane that performs the same function as the film in the camera and is characterized by being very thin. It contains ten layers of nerve cells, nerve fibers, photoreceptor cells, and support tissue. The retina works by converting light rays into nerve impulses that are transmitted through the optic nerve to the centers of the brain to translate these signals into sensible images.

10.2.6.2 Retinal blood vessels
The external layers of the retina (up to the noticeable pterygium layer) are taken care of by a component that pervades the vascular layer of the placenta, while the inward layers are taken care of by the central retinal vessels. The retina does not have tangible innervations, so its issue is effortless. The corridors travel in a radial manner, dissimilar to the veins that move in an alluring attraction.

10.2.6.3 Central artery of the retina
The first branch of the ophthalmic artery is a part of the internal carotid artery, a valid and terminal corridor related to a vein and a thin course that separates into two upper and lower trunks, typically at the optic disk. The optic nerve enters approximately 1 cm behind the chunk of the eye, around 7 mm. A blood vessel beat happens if the circulatory strain is lower than the blood vessel pressure inside the eye. The tissue comprises an endothelial layer made of a solitary layer of endocrine cells, the endothelial restricting layer, the center layer consisting of free connective tissue, and the epithelial layer containing free connective tissue [4].

The Windkessel model, designed in the late 1800s by German physicist Otto Frank, describes the heart and the systemic arterial system as a closed hydraulic circuit. According to this view, the circuit contains a water pump connected to a water-filled chamber except for an air pocket, in which blood flow through arterioles is described as the flow of water through tubes. The Navier—Stokes equations, developed by scientists Claude-Louis Navier and Gabriel Stokes, are non-straight halfway differential conditions that portray the development of fluid materials and were used in describing the flow of blood through the human body [5]. Scientist Bear developed a Fabry Perot transponder, which is the simplest laser resonator consisting of a pair of flat mirrors or circular mirrors that are positioned opposite and centered on a single axis, which have been presented in this project to simulate the generation of a laser beam. Dermats and Falchus additionally built up a calculation for vascular discovery called the streak-following strategy used to follow the line on the picture utilizing a particular rakish direction and vessel distance across by taking advantage of the histogram distributed image. The pixel area boundaries are defined to be tracked by the threshold value corresponding to the image density frequency. After obtaining a tracking region, initialization is done to track neighbors' pixels one by one by direction and their predetermined diameter [6]. By calculating the weight value of each neighboring pixel, the pixels with the largest weight and exceeding the predetermined threshold value are determined. Chanwimalang created a dialog to detect the vascular efficiency of the retina and investigate clots in it using a threshold level of entropy. It is an effective method for the automatic detection and extraction of retinal vascular images. The proposed algorithm consists of four steps: filtration matching, local entropy threshold level, filtration length, and vascular junction detection. This technique produces a two-dimensional vector. The techniques for measuring blood flow are Doppler or 3D laser. The Doppler laser is used to spot cases that may evolve into situations that contain high concentrations of particles. That is why research has spread using the Doppler laser to detect blood clots [7].

10.3 Doppler laser

The retinal blood flow velocity meter is based on projecting a beam of light through a guiding system onto a blood vessel in the retina and forming a separate tracking image through the aiming system. A fast-track loop checks the motion of the tracking image and moves the guiding system to cancel the motion of the image and keep the light beam focused on the bowl.

The beam is reflected from the vessel so that the detectors capture it at two fixed angles to be processed by spectroscopy. In one preferred inclusion, the optical beam and the tracking image follow two separate paths in the directed system [8]. Optical fibers transmit the collected Doppler light without scattering it, while the phase relationships and absolute dimensions are preserved electronically by determining the tracking image. The therapist then calculates the volumetric flow of blood, which is compared with standard data. The current technology is linked to machines that measure blood flow in the vessels of the retina through Doppler meters to measure flow velocity.

The general theory of laser Doppler flowmeters, as applied to flow field measurements, for example, the blood within a blood vessel, is a well-known application of flow measurement techniques. In short, the monochromatic light directed at the vessel and into the circulating blood is reflected by the blood cells in the form of diffused light, which is accompanied by a frequency distribution corresponding to the velocity components of the individual scatterers. By analyzing the frequency distribution of the reflected light at two fixed-angle receivers equipped with a known separation angle, the velocity or form is determined. An ideal profile/gravity picture of the speed of the running blood can be inferred [9].

10.3.1 Doppler analyzer

The stable laser Doppler system 10 includes a 20-point guidance kit, a tracking group of 30 red laser sources 35 to illuminate the retina vessel, a two-channel Doppler imaging camera, and a 40-analysis set. The guidance group, which includes mirrors inclined on two axes x and y, is controlled by electronic signals to direct the optical path of the beam from the red laser to the desired retinal vessel, and then the beam remains centered on the target vessel by the tracking system, which in turn controls the position of the image of the same blood–retinal vessel. Image position changes are verified on the CCD array and used to develop control signals that reposition the steering mirrors to avoid image movement [10]. These techniques are described and are aimed at deriving well-defined tracking signals and controlling the mirrors quickly and accurately enough to direct the beam from the 35 lasers at the target retina. The red laser light scattered from the target retinal vessel in the back of the eye is imaged by the target optics in the eye returning toward the guidance system as it is deflected by a pair of 42, 41 mirrors, each containing a diameter of about

1 mm, and a distance of 6 mm from each other. The viewing light is supplied to the system by a yellow helium-neon laser 51 and is directed through an expanding beam 52 passing through the diffraction mirrors a53, b53 and through two modes of light 54 to provide a wide field beam, which is multiplied within the optical path by a mirror that divides the light beam into several paths 55. The ray is illuminated about 10 degrees from the bottom. The yellow viewing light reflected from the bottom through the target group that contains lenses also contains an image manager and passes through the steering group to the lens, where it provides a visible field of view that moves simultaneously with the tracking image and with the target bowl and the Doppler light spot. The vision group can contain a camera and the function of the image manager, which is simply to rotate the tracking image (such as a retinal vessel), for example, a fixed frame perpendicular to the CCD [11].

The Doppler illumination beam emitted by the laser is focused on a speck of the vessel approximately equal to the diameter of the target vessel of the retina, and the energy of the incident beam is reduced to about 5 microwatts. Consequently, it reaches the biologically permissible safety limit for the exposure of the retina to the intensity of the radiation flow. While the multiplexer tubes necessary to verify the reflection of radiation at these low levels may escape the radiation state due to its natural approval of stray reflections [12].

10.3.2 Doppler imaging models

A retinal vessel may contain a width of under 50 to a few 100 μm. It is outlined by a green 90-shading tracker bar, which has a roundabout or rectangular cross-segment with a distance across 5−10 mm, and the Doppler light bar is centered to spot 95 around a similar vessel. The retinal region enlightened by bar 90 is pictured and adjusted by the direction framework 20, as portrayed in picture 90 on a 130-CCD straight grid arranged opposite to picture 120 on the vessel. An enhancement blend of 5−15 intensifiers is utilized so that the CCD lines are totally situated inside the 90 pictures of the following shaft. For instance, a straight cluster comprising one increased by 256 pixels (CCD), which is approximately 12 mm in length, and 25 energy targets, which guarantee that the picture of a shaft followed by a width of 500 μm, will cover the 130 CCD. Similarly, a 50 to 100 μm picture of the retinal veins will cover around 25 to 50 pixels of the CCD [13].

10.4 Proposed methods

10.4.1 Scientific method

In this study, we present an accurate and fast algorithm to verify blood clots and measure blood vessels in retinal images through two main steps. The first step is a simple approach that uses algorithms in the MATLAB® language. It divides the vessel by augmentation and wavelet parameters. It is portrayed as being more rapid than the connected component algorithm based on images to extract the axes and position the edges of the vessels from the image profiles by using a suitable chip to determine the vessel orientations and then search. The zero crosses the crossing into the second derivative of the line perpendicular to the bowl, using the bottom images included in the standard database and diameter extraction measurements from the detected edges. We also confirm the analytical solutions found by Windkessel in modeling blood flow during the cardiac cycle using MATLAB as well as a simulation of a helium laser using resonance medium equations. After being reflected from the retina, its reception is simulated on the CCD matrix, and Doppler shifts are detected in the received signal on the CCD matrix [14].

10.4.2 Statistical analysis

10.4.2.1 Analysis and simulation of blood flow using MATLAB

Reproducing the progression of blood through the human body is basic in the medical field. Reenacting this normal stream could give specialists data about how the medications work in the body. This method can help analysts plan precise treatments that will improve patients' wellbeing. In order to do this, a restricted contrast technique is used, which entails evaluating vein reconstruction using MATLAB through unidirectional resistance modelling.

10.4.2.2 Rotation system

It is important to become familiar with the rotational framework for simulating fluid flow through mathematical expressions before diving into any numerical conditions. Blood has two components: cells and plasma. Platelets skim in plasma is the fluid bit of blood. The segments are the supplements, hormones, and the natural protein inside the plasma. Plasma diffuses these substances as they circle through the body.

10.4.2.3 Number and senders

The Womersley number is a dimensionless number that confers fluid mechanics and arrangements to the pulse-flow frequency. It comes to address Navier–Stokes' conditions. It shows the proportion of the oscillatory idleness power to the shortening power. In a vascular retina, parameters, such as dynamic frequency, density, and viscosity remain the same, except for tube (vessel) radii [15].

Reynolds numbers are significant in bigger veins and less so in smaller ones. Given that blood's density and viscosity are generally consistent, the square root estimation is true in any situation. As a result, the number of blood vessels becomes an important aspect to take into account.

10.4.2.4 Navier–Stokes equations

The Navier–Stokes conditions are nonlinear halfway differential conditions that portray the movement of liquid materials. These conditions come from applying Newton's second law to smooth movement. Newton's subsequent law expresses that the pace of progress of energy is relative to the force and proceeds toward the force. The Navier-Stokes equations are useful in a variety of domains, including the design of vehicles and aircraft, the analysis of environmental pollution, and the regulation of fluid velocity at particular geographical locations. This is in addition to their use in the research of blood flow. At the point when speed is discovered, different amounts of stream rate can be calculated. Isaac Newton's subsequent law (upkeep of energy) shows that power approaches mass duplicated by quickening; F = mama. It is clear from Newton's second law that an object's mass and net force act as the main influences on velocity. In the context of fluid dynamics, it becomes evident that contact is a result of fluid resistance and fluid pressure, with grinding acting in contrast to acceleration. The power produced by the liquid and the mass of the liquid are considered. By applying the ideas of Newton's second law to the movement of liquids, it is simpler to portray the properties of acids, for example, blood, in scientific terms [16].

The two conditions of the Navier–Stokes can be written in part as follows:

$$\rho\frac{\delta u}{\delta t} + \rho\left(\frac{\delta\left(u^2\right)}{\delta x} + \frac{\delta(uv)}{\delta y}\right) - \nu\left(\frac{\delta^2 u}{\delta x^2} + \frac{\delta^2(u)}{\delta y^2}\right) + \frac{\delta\rho}{\delta x} = f1$$

$$\rho\frac{\delta u}{\delta t} + \rho\left(\frac{\delta(uv)}{\delta x} + \frac{\delta(v^2)}{\delta y}\right) - v\left(\frac{\delta^2 v}{\delta x^2} + \frac{\delta^2(v)}{\delta y^2}\right) + \frac{\delta\rho}{\delta y} = f2$$

$$\frac{\delta u}{\delta x} + \frac{\delta v}{\delta y} = 0$$

where, ρ = Thickness (3 cm/g), V = Liquids are viscous (g /(cm^2 * sec)), P = p(t,y,x), that is, pressure at the point (x,y) in time t = (g/(cm * sec^2)), F = any outer power following up on the fluid, and $\mathscr{F} = (\mathscr{F}_1(x, y, t), \mathscr{F}_2 \ (x, y, t))T$ [17].

10.4.2.5 Finite difference method
For displaying the exact bloodstream, the limited distinction technique is utilized. Navier–Stokes conditions enable an arrangement of nonlinear incomplete differential conditions for the bloodstream and the cross-sectional territory of the supply route. Here, the limited distinction technique is utilized to settle the conditions mathematically; the utilization of these outcomes leads to the option of screening the most precise changes in the bloodstream. This strategy does not involve finance, which makes the proposed system cost-effective. Leveraging computational analysis's differentiation techniques to estimate differential problem solutions. To use this method of resolving issues the domain must be divided into a closed system over the time period.

10.4.2.6 Numerical chart
To reenact one part of the bloodstream, we assessed the stream rate (q) and the cross-sectional region (A) regarding a segment of focus stretching out along the branch and then rehash this level from the primary level to the subsequent level. Furthermore, in some parts of the focus, we will have the example size (k) and we will save this number as 100. Although, the higher the example size, the more precise the estimation of the outcomes. There are a couple of conditions that are basic in making this reproduction one-dimensional. The equations of flow and weight can be executed by saving mass and pushing as follows:

$$\frac{\partial q}{\partial t} + \frac{\partial q}{\partial x} = 0$$

$$\frac{\partial q}{\partial t} + \frac{\partial q}{\partial x}\left(\frac{4}{3}\frac{q^2}{A}\right) + \frac{A}{\rho}\frac{E_{stat}h_0}{R_0A_0}\frac{\partial A}{\partial x} = 8\pi v\frac{q}{A} + v\frac{\partial^2 q}{\partial x^2}$$

where A = cross-sectional area, q = flow rate, ρ = density, Labs ESTAT = young consistent, h = wall thickness, R_0 = the radius of the original ship, h_0 = the original wall thickness, and A_0 = area of the original sectional [18].

As a result of using the Doppler analysis unit to reflect light from the outer surface of the blood vessel as a reference ray, the effects of the frequency shift caused by eye movement or light spot, and flow rate are the same whether the center Doppler beam is stationary or moving along the direction of the blood vessels. For this reason, the tracking system does not require a controlled movement in two directions, but it can be a one-way tracker with its tracking components to perform the correction process in a directed fashion only for eye movement in a direction that intersects with the vessels at which the Doppler beam is directed.

10.4.2.7 Processing workflow

First, the processor completes the axis flow speed for a period of time, through numerical processing, and divides the integral to determine the average speed of the axis blood. Second, the processor determines the cross-sectional area of the vessels according to: (A = π (d/2) 2) from the vessels' diameter for large enough vessels (greater than 50 m) [19].

10.4.2.8 CCD camera

A superior method of noticing a point of interest in a laser shaft is by making a variety of finders. For waves in the visible area and close to infrared beams, the CCD camera can be utilized as a sensor cluster. Electro-warm clusters put by lasers that work out in the areas in which the CCD camera is delicate (NM011–nm111) with the fact that the standard camera sensors are exceptionally touchy to light, and laser light requires weakening before it arrives at the sensor. This can be halted using clear mirrors that permit a small amount of the laser to go through. Impartial thickness channels can also be utilized to hose the light. Mirror channels are gathered, so it is essential to adjust them to a layered exchange to lessen obstruction phenomena that can misshape the radiation picture. Residue on the mirror channels can prompt genuine contortion of the laser bar picture. To play out the laser bar estimation in this task, the camera strategy was picked, and the principal purpose behind this was to impart the full bar mode with 2D innovation and to make snappy estimations with little vision through the camera. The laser heartbeats can be isolated and measured. With a CCD camera, the total picture of the laser shaft is taken, at the same time, making it conceivable to gauge single heartbeats from a beat light and a laser switch. From that point, the camera gives the picture of the shaft force, the picture is prepared, and the width and laser energy are removed and afterward isolated.

10.4.2.9 Calculation methods
The purpose of the task is to screen a pillar and give a warning in the event of changes. In the event that the laser position is changed, the bar width will also be altered. The complete energy of a pillar will coordinate the absolute thickness. It is hard to quantify the total energy since all energy attenuators have to be aligned. It is smarter to gauge the force by different methods to quantify the overall strength with the camera and caution the administrator when the energy changes and the beam width is characterized as multiple times the standard deviation of the shaft. Other ways to define a package width are full width at half the greatest thickness limit [20−22].

10.4.2.10 Variance
The ISO 11146 standard indicates a shaft width multiple times the standard deviation. In the event that a beam is rounded or curved and is lined up with the x and y tomahawks, the beam can be shown on the x and y tomahawks, and the beam width can be determined from the projections.

This disentangles the figuring cycle, as it will give a bogus width if the home is pivoted to get the right width. The basic difference between x and y is utilized. The width is determined by the oval hub. The changes from the projections demonstrate the tomahawks axes ($<y2>$ and $<x2>$), and the common variance of the beams is $<xy>$. These are calculated using the following three equations:

$$<x^2> = \frac{1}{p}\int (E(x,y).(x-\langle x\rangle))^2 dxdy$$

$$<xy\frac{1}{p}\int \left(E(x,y).(x-\langle x\rangle)(y-\langle y\rangle)\right)dxdy$$

$$< y^2 > = \frac{1}{p} \int \left(E(x, y).(x - \langle x \rangle)(y - \langle y \rangle) \right) dx dy$$

The beam width closest to the X-axis is calculated using:

$$d_{\sigma x} = 2\sqrt{2} \left\{ \left(\langle x^2 \rangle + \langle y^2 \rangle + y \right) \left[\langle x^2 \rangle - \langle y^2 \rangle + 4(xy) \right]^{\frac{1}{2}} \right\}^{\frac{1}{2}} \dots [21].$$

10.5 Results and discussion

As the beat moves between 120 mmHg and 80 mmHg during the cardiac cycle, the model is also able to absorb the changes in the blood components during the cardiovascular cycle as it was for the two-component model. One can observe the analytical solution corresponding to the numerical solutions computed using MATLAB. However, the validity of the model cannot be proven more than that in the absence of patient data. The blood resistance value (modeled in the three-element model) is neglected when compared with the peripheral resistance. In the proposed algorithm the labels of the vessels are sorted according to the segment length or mean diameter. This makes it easier to identify and remove slices with extreme measurements, which appear as numbers in the container. The location of the optic disk and displaying circles to show the peripheral areas in the optic disk can be manually determined by the retina. This is useful for some protocols where it is only necessary to measure within specific distances from the optic disk. It is also useful for determining and separating blood clots.

10.5.1 Results of application of the algorithm to several images

10.5.1.1 Dark vessels

During the segmentation phase you need to determine whether the vessels are darker than the background (fundus image) or lighter (cardiac fluorescence). This adjustment affects whether the vessels are detected at all (fully), although their value is usually determined and correction is easy.

10.5.1.2 Wavelet levels

These should be set according to the resolution and size of the vessels to be detected. Large blood vessels require higher levels than the numbers they indicate. The preview image shows how the output will be for different options.

10.5.1.3 Connectivity restrictions

The edges are revealed as there is a sharp gradient around the bowl. Most of the time the steepest gradations near the vessels correspond with their true edges, but sometimes they may belong to other vessels. Many of the limitations that already exist in the detection algorithm are incorporated, but these may not be sufficient to ensure that the correct edges are found, and the connection constraint forces the algorithm to accept only large gradients that form a long continuous line close to the vessel and thus ignore most other pseudo gradients, which are usually shorter. Fig. 10.1 shows the stability diagram for laser.

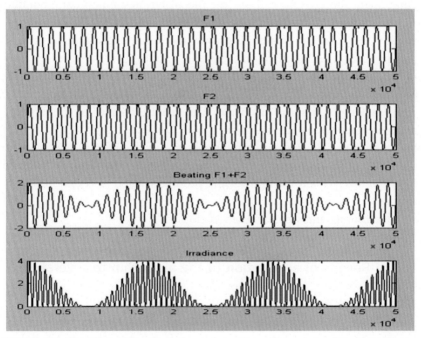

Figure 10.1 The phenomenon of the pulse of two frequencies (F1 and F2); the detector responds to the radiation (MATLAB graph), which due to its response time produces a cover that covers the curve.

The blue point must be inside the stability zone defined by the red curves and lines ($0 = g2$, $0 = g$). If it is outside the field, then the resonance medium will be unconfined, and therefore no pattern can be found and the laser is outside the stability zone.

10.5.1.4 Simulation of the pulsating frequency phenomenon

By using the general principle of the pulsating frequency phenomenon and the self-mixing theory, the results shown in the figure were reached, where a one-dimensional beam was sent to the medium and the returned beam was monitored. Mathematical equations that represent the mixing phenomenon were used. With these results, information about the target object and the laser itself can be obtained.

10.6 Conclusions

The extension of the automatic analysis of the retinal images (p) is due to the nonlinear performance of the methods used. This problem has led to some obstacles in clinical detection cases. Therefore, the algorithm must be developed so that the vessel detection method uses filters that have a circular profile, to accelerate detection of the next point along the vessel. To expand the use of this algorithm, work is being done to develop programs capable of separately treating the vessel extracted from the retina, and determining the dimensions of the thrombus within the blood vessel with the greatest accuracy, and using the grayscale of the vessel as an indication of the pathology. Future work is targeted at developing more accurate algorithms to process the signal reflected from the moving particles in the blood vessel to detect the presence of a thrombus and improving the durability of this system through the addition of a module capable of separating the anatomical parameters associated with the vascular tree. The system will be developed to obtain high-quality, real-time continuous indication that describes real-time blood flow. By achieving the aforementioned, the doctor's needs are met in assessing vascular function, with special emphasis on the state of flow and their ability to predict diseases that may occur in patients.

References

[1] A. Dance, Blood Clots: A Major Problem in Severe Covid-19, Knowable Magazine, 2020.

[2] B. Baumann, B. Potsaid, M.F. Kraus, J.J. Liu, D. Huang, J. Hornegger, Total retinal blood flow measurement with ultrahigh speed swept source/fourier domain OCT, Biomedical Optics Express. 2 (2011) 1539−1552.

[3] C. Delaey, J. Van De Voorde, Regulatory mechanisms in the retinal and choroidal circulation, Ophthalmic Research 32 (2000) 249−256.

[4] B.L. Justesen, P. Mistry, N. Chaturvedi, S.A. Thom, N. Witt, D. Kohler, Retinal arterioles have impaired reactivity to hyperoxia in type 1 diabetes, Acta Ophthalmologica 88 (2010) 453−457.

[5] P. Acevedo Tapia, C. Amrouche, C. Conca, A. Ghosh, Stokes and Navier-Stokes equations with Navier boundary conditions, Journal of Differential Equations 285 (2021) 258−320.

[6] B. Haney, A proof for navier-stokes smoothness, SSRN Electronic Journal (2021).

[7] V. Doblhoff-Dier, L. Schmetterer, W. Vilser, G. Garh€ofer, M. Greoschl, R.A. Leitgeb, et al., Measurement of the total retinal blood flow using dual beam Fourier-domain Doppler optical coherence tomography with orthogonal detection planes, Biomedical Optics Express 5 (2014) 630−642.

[8] J.P. Garcia, P.T. Garcia Jr, R.B. Rosen, Retinal blood flow in the normal human eye using the canon laser blood flow meter, Ophthalmic Research 34 (2002) 295−299.

[9] W. Li, J. Zhou, Novel laser doppler tachometer, Chinese Optics Letters 19 (1) (2021) 011201.

[10] A. Unewisse, Doppler characteristics of HF spread doppler clutter recorded at alice springs, Australia, Advances in Space Research 67 (3) (2021) 1026−1038.

[11] W. Tianthong, V. Phupong, Serum hypoxia-inducible factor-1α and uterine artery Doppler ultrasound during the first trimester for prediction of preeclampsia, Scientific Reports 11 (1) (2021) 6674.

[12] D. Melber, J. Ballas, Clinical applications for doppler ultrasonography in obstetrics, Current Radiology Reports 9 (2) (2021) 1−10.

[13] A. Leitão Ferreira, Laser doppler flowmetry, Universidade de Coimbra, Graduation in Biomedical Engineering, Faculdade de Ciências e Tecnologia da Universidade, 2007, 1−66. 4, pp. 72−79.

[14] D.A.L. Paul, Bailey. D., The Helium-Neon Laser, University of Toronto, advanced undergraduate laboratory, 2011, pp.1−11.

[15] E. Ingemar, The Monitoring of a Laser Beam, Department of Information Technology and Media (ITM), Master, Sweden, Mid Sweden University, 2005, pp. 1−43.

[16] E. Siegman, Measure Laser Beam Quality, Tutorial presentation Annual Meeting Long Beach, at the Optical Society of America, California, 1997.

[17] F.F.M. de Mul, M.H. Koelink, A.L. Weijers, J. Greve, J.G. Aarnoudse, R. Graaff, et al., Self-mixing laser doppler velocimetry of liquid flow and of blood perfusion in tissue, Applied Optics 31 (1992) 5844−5851.

[18] H. Jukka, Self-mixing interferometry and its applications in noninvasive pulse Detection, Department of Electrical and Information Engineering, University of Oulu, Finland (2003) 1−74.

[19] M. Catanho, M. Sinha., V. Rsha, Model of Aortic Blood Flow Using the Windkesseleect, Mathematical Methods in Bioengineering (BENG), 221, 2012, pp.1−16.

[20] P. Bankhead, Automated retinal image analyzer v1.0, centre for vision and vascular science, Queen's University of Belfast, UK, Nikon Imaging Center @ Heidelberg University, Germany, SIS-ISO/TR 11146−3:2004, 2011.

[21] T.. Milbocker, Retinal Laser Doppler Apparatus Having Eye Tracking System, Eye Research Institute of Retina, Eye Research Institute of Retina, Patent (5,106,184), Boston, 1992, pp. 1−15.

[22] Z. Christian, M. Dickinson, T. King, Dynamic light scattering by using self-mixing interferometry with a laser diode, Applied Optics 45 (10) (2006) 2240−2245.

CHAPTER ELEVEN

Sleep apnea detection device

Dilber Uzun Ozsahin[1,2,3], Basil Bartholomew Duwa[3,4], Bartholomew Idoko[5], Ayman Aleter[4], John Bush Idoko[6] and Ilker Ozsahin[3,7]

[1]Department of Medical Diagnostic Imaging, College of Health Science, University of Sharjah, Sharjah, United Arab Emirates
[2]Research Institute for Medical and Health Sciences, University of Sharjah, Sharjah, United Arab Emirates
[3]Operational Research Center in Healthcare, Near East University, Nicosia/TRNC, Mersin 10, Turkey
[4]Department of Biomedical Engineering, Near East University, Nicosia/TRNC, Mersin 10, Turkey
[5]Department of Physics, Advance Educational Consult and Research Institute, Abuja, Nigeria
[6]Applied Artificial Intelligence Research Center, Department of Computer Engineering, Near East University, Nicosia/TRNC, Mersin 10, Turkey
[7]Department of Radiology, Brain Health Imaging Institute, Weill Cornell Medicine, New York, NY, United States

Contents

Practical Design and Applications of Medical Devices.
DOI: https://doi.org/10.1016/B978-0-443-14133-1.00008-2
147

11.1 Introduction

Sleep apnea diseases are recorded as one of the most prevalent diseases experienced by many individuals due to either psychological interferences or physiological appearances. Sleep apnea is one of the most common diseases nowadays, and its signs include annoying snoring and choking during sleep. Even though the disease causes sleep disturbance at night and drowsiness, lethargy, and poor concentration during the day, it has many organic complications for important body organs such as the heart and brain. The disease results from the narrowing of the upper airway (pharynx) during sleep. This results in recurrent partial or complete blockage of the throat. This causes a severe and frequent lack of oxygen in the blood, which leads to the release of oxidizing agents in the blood, raises blood pressure, irritates the nervous system from tension, and releases stress hormones, such as adrenaline and cortisone in large quantities during sleep. This is contrary to the nature of the body, as the level of these hormones decreases during sleep. In normal people, the nervous system calms down, which leads to heart and brain relief, and decreased heart rate and pressure. The cessation of breathing in a normal person does not exceed five times an hour, and the cessation of breathing is considered to be 5–10 times per hour light and from 16–30 times per hour on average if it is more than 30 times an hour, it is considered severe. Organic changes resulting from the repeated cessation of breathing during sleep lead to atherosclerosis. This condition induces repetitive full or partial throat blockage, leading to extreme and regular loss of cholesterol in the body, which results in the release of oxidants into the slot, raising blood pressure, irritating nerves from strain, and releasing stress hormones, such as adrenaline and corticosteroids. Breathing stoppages in a normal person occur not more than five times an hour. In sleep apnea, breathing stoppages are calculated to be 5−10 times an hour and, on average, 16−30 times per hour, which is more than 30 times per hour. A lot of recent research has shown that there is a strong link between sleep apnea and brain strokes [1−3]. Diagnosing sleep apnea is not easy. The reason is that different reasons lead to an increase in sleepiness during the day, and the patient needs to conduct a study to diagnose the disease and determine its severity.

The primary function of the respiratory system is to deliver oxygen into the blood and get rid of carbon dioxide. First, the knowledge of the components of the respiratory system that enable it to function properly will be elaborated. The respiratory system starts from the nostril, the pharynx cavity, the larynx, the trachea, and then the bronchi to the alveoli, and each part has a specific characteristic. Fig. 11.1 shows the main parts of the respiratory system.

11.2 Components of the respiratory system

11.2.1 Nose

The visible part of the nose consists of the cartilage and a bony part. The nasal cavity is divided from the inside into two parts, the nasal septum and the nasal cavity, which begins in the front with the two front nostrils and ends in the back with the two posterior nostrils that open into the throat. The nasal cavity is also lined with a mucous membrane, which contains a large number of capillaries and mucous glands that secrete a mucous substance, lubricating the cavity. At the two front nostrils there is a small amount of hair to trap foreign bodies and dust particles from the inhaled air. The mucous membrane lining

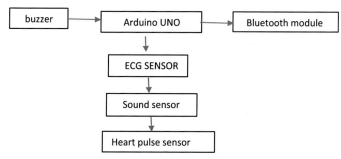

Figure 11.1 Device design.

in the nasal cavity is fed by several nerves, some of which are olfactory nerves in the upper part and some sensory nerves in the lower part.

11.2.2 Pharynx

The pharynx is a tube of muscle that is wide at the top and narrow at the bottom and extends from the base of the skull to the sixth cervical vertebra. It is about 14 cm long, and the wall of the pharynx is made of muscles that line it from the inside as a mucous membrane.

11.2.3 Larynx

The larynx is also called the voice box, because it includes the vocal cords, which are connected to the trachea. The larynx is the last part of the upper respiratory tract and consists of rings that connect through muscles and membranes. The most important part of the larynx is the cartilage of the thyroid gland, and it forms what is called Adam's apple. Sensory nerves and fibers help to provide protective mechanisms, such as coughing.

11.2.4 Trachea

Some believe that the windpipe is only a tube for air to pass into the lung, but the trachea has a structure that enables it to perform a specific function. It is a cylindrical tube that extends from the bottom [2]. The larynx reaches the fifth thoracic vertebra and is 53 cm long. Its front surface is convex, and its posterior surface is flat, nearly where it comes into contact with the esophagus.

The tracheal wall consists of many cartilages, but these cartilages only cover the anterior part [3]. The trachea to the posterior part of the wall is made of muscles, not cartilage, and this formation (the windpipe) is allowed to be solid and open to the sky to pass through the air and at the same time give its flexibility as it allows the muscular part to contract. This characteristic is necessary for two important functions:

1. Issuing different sounds where the constriction of the windpipe is necessary to create a stream of air. The outside of the lung enables the vocal cords to produce sound.
2. Coughing is a bit annoying, but it has an important benefit in helping a person. As depicted in Fig. 11.2, the cough gets rid of phlegm or harmful secretions that might form in the lung, enabling flexible airways for the person if they cough effectively.

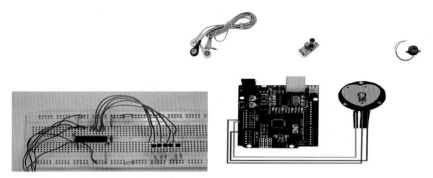

Figure 11.2 Set-up of the device.

11.2.5 Bronchioles

After the windpipe branches into a left and right part, these tubes gradually divide to form a network. The tube's function is to deliver air to various parts of the lungs, and these bronchioles are very important, as they remain open to the air during the inhalation and exhalation process. But in some cases, such as bronchial asthma, the airway in these bronchioles narrows, and this narrowing is the main reason for shortness of breath and wheezing experienced by asthma patients [1].

11.2.6 Alveoli

The lungs contain approximately 300 million air vesicles known as alveoli, which is surrounded by a delicate network of large capillaries. The overlap and coordination between the air coming from the outside atmosphere with the oxygen and blood coming from the heart laden with carbon dioxide allows oxygen to move from the alveoli into the capillaries, thus transporting it to all parts of the body while at the same time getting rid of carbon dioxide [1].

11.2.7 Lungs

The lung contains arteries, pulmonary veins, and nerves located in the chest cavity. The lung is almost pyramidal, with a top pointing up and a base pointing down, and each lung has two surfaces and three corners. The lung apex extends down the neck and above the collarbone. The lung base rests on the diaphragm muscle, separating the chest cavity from the abdominal cavity, and the base of the lung is concave. The lungs are much larger than most people think. They extend from the neck down to the diaphragm. It is the separator that divides the body cavity into two parts, which roughly resemble a cone, its top up and its base down, and each lung is independent of the other, such that if you cannot breathe with one for any reason, the other one becomes active for use [3]. The left lung consists of two lobes, while the right lung consists of three lobes. The lung is spongy from the inside, lightweight, and contains millions of air sacs. The lung is very elastic and can expand easily, but once the force disappears, it stretches until it returns to its original shape [4−7].

11.2.8 Respiratory muscles

The respiratory muscles are the diaphragm muscle and the muscles of the chest and abdomen, and breathing is greatly dependent on the flexibility of these muscles and their ability to move the chest cavity appropriately [5].

11.2.9 Diaphragm

The diaphragm is a dome-shaped muscle located at the bottom of the lungs. It is used for breathing as it separates the chest cavity from the abdominal cavity [7].

11.2.10 Rib cage

The chest cavity is like a tight box that contains the lungs and the heart. The ribs surrounding the rib cage are placed on the top and sides, while the diaphragm is located at the bottom [6].

11.3 Physiology of respiratory system

11.3.1 Breathing

Breathing is different from ventilation. Ventilation is simply the entry and exit of air from and to the lungs. This helps in the propagation of oxygen inside and carbon dioxide out of the lungs [4]. The actual breathing takes place in two phases:

1. First phase: Within the lungs, where gas exchange takes place between the blood and the air entering the lungs, i.e., the transfer of oxygen from the air to the blood and carbon dioxide from the blood to the air.
2. Second phase: It is the opposite of the first phase, but it is done at the cellular level, i.e., between cells and blood.

11.3.2 Respiratory regulation

Inhaled air contains 21% oxygen and about 0.03% carbon dioxide, in addition to many gaseous waste. Respiratory regulation is essential to keep oxygen and carbon dioxide levels constant in the blood; the nervous system controls this process [8−11]. The amazing thing about breathing is that we do not that we have to think about it, whether we are asleep or awake, because breathing occurs automatically, as the central control of breathing in the brain does this job. In the blood there are sensor-like components that are sensitive to oxygen and carbon dioxide levels in the blood, sending messages to the respiratory control center. The brain controls these levels, for example, if the level of oxygen in the blood is less than the permissible limit, the brain stimulates the lungs to increase the respiratory rate to compensate for the lack of oxygen. It has been found that a person's breathing rate ranges between 12 and 20 breaths per minute.

11.4 Causes and diagnosis of sleep apnea

11.4.1 Inhalation

Inhalation is a vital process. For an inhalation to occur, the diaphragm contracts and decreases further from the volume of the chest cavity, creating a negative pressure (pressure drop) within the cavity, and making air enter the lungs. As the air moves from the high-pressure region to the pressure zone, the chest muscles also raise the ribs so that the chest cavity expands from the front, as shown Fig. 11.3.

11.4.2 Exhalation

Exhalation is a passive process, and as such, there are no primary muscles involved in exhalation. As the diaphragm and external chest muscles relax, no muscular effort is required for exhalation to take place [12]. It may be possible to recruit auxiliary muscles (the inner intercostal muscles in the abdomen) to help exhalation occur when the airway is blocked, during periods of exercise, coughing, or sneezing.

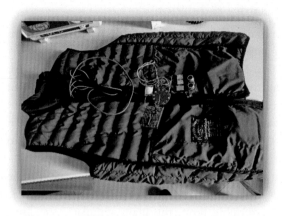

Figure 11.3 Prototype.

 ## 11.5 Types of sleep apnea

11.5.1 Obstructive sleep apnea

Obstructive sleep apnea (OSA) is caused by the relaxation of the muscle mass in the neck, causing the airway to close. It is the most common type of sleep apnea present in adults, with about 5% in males and less in females before menopause. The percentage in women after menopause is equal to that of men. It is usually experienced when the airway closes but the breathing movements continue. It is generally common in people over the age of 50 and is more common in men than in women.

11.5.2 Central sleep apnea

Central sleep apnea (CSA) is caused by a defect in the brain's breathing control center. It is uncommon. It occurs when breathing stops even though the airway is still open. It occurs when a person loses the ability to breathe spontaneously, and this results from a disorder in the central nervous system. It can be a result of some diseases, such as some brain diseases (stroke, brain tumors, and viral encephalitis), or a chronic disease of the respiratory system. It can be rarely associated with snoring, and, as such, can be difficult to diagnose.

11.5.3 Mixed sleep apnea

Mixed sleep apnea is caused by a defect in the breathing control center in the brain, in addition to a relaxation of the muscles of the neck. It is the least common and is a combination of OSA and CSA. The person with this type suffers from snoring, and therefore the treatment helps to get rid of obstruction of the airway, but it is difficult to eliminate sleep apnea. The treatment is a combination of OSA and CSA therapy.

11.6 Diagnosis of sleep apnea

The diagnosis must be undertaken by treating professionals (physicists, pulmonologists, specialists in neurology, and physicians trained in sleep disorders). It is difficult to diagnose because there are many causes of sleep apnea, and the diagnosis must include a medical history and a complete physical examination. Polysomnography is the most common test used to determine whether a person has sleep apnea. Sometimes the simplest diagnostic procedures can identify the disease. However, if the test does not confirm whether a person is infected, polysomnography should be PERFORMED [10].

11.7 Methods and device performance

This device was designed to investigate and monitor the breathing sound as a result of apnea. Collective comparisons from different studies on apnea guided the methodology and approach for this development. Based on the descriptive studies at our disposal, apnea could be in three phases, mild, moderate, and severe. Constructively, severe and moderate apnea are the riskiest forms of apnea, while mild apnea is less harmful. Furthermore, our knowledge of the different forms of apnea has assisted us in taking a decisive approach to the construction of this ambient device. The device was designed based on the studies of Dhruba et al. [11].

The block diagram consists of the components of the device, such as the input and output connections, microcontroller board Arduino Uno, sensors, and a mobile app. The device incorporated components that analyze the tracheal sound waves and subsequently detect sleep apnea above the normal frequency of about 400 and 700 Hz. If sleep apnea is detected, then the device makes sounds based on the program.

The components utilized for the construction of the proposed novel device include a power supply, a block diagram of the feeding unit, a calendar bridge for the full router, voltage regulators, a microcontroller, characterization of the microcontroller terminals, a digital-analog switch, microcontroller action, an infrared sensor, the dispatch section, the department of the future, a nose sensor and a snoring sensor, a microphone, stress microphones, capacitive microphones, frequency response, an LCD screen, and a thermal sensor. We interconnected these components to build the proposed system.

We used the AVR8-bit microcontroller (ATMEGA), based on the RISC (reduced instruction set computer) technology, as the kernel of the system for the control. We opted for the ATMEGA microcontroller because it has the capacity to work under low frequencies and can scale comparative patterns [4]. The microcontroller receives data from both the snort sensor, the incoming air (from both mouth and nose) sensor, and the temperature sensor to determine whether there is snoring followed by a pause in breathing [13–15]. If this condition is achieved and the patient does not wake up within a maximum of 51 minutes, the controller automatically activates the buzzer to generate a sound (alert) to wake the patient from sleep, adopting a similar technique used by Ravi and colleagues [4]

11.8 Components

11.8.1 Arduino microcontroller

Arduino is one of the major components used in the design of the sleep apnea detection device. The Arduino board contains numerous connections with different digital pin inputs. The Arduino Uno model is the central connector among the various components. When the Arduino Uno is connected to the components, the instruments become more viable and active in detecting an anomaly in sound regarding apnea.

11.8.2 Buzzer module (piezo speaker)

The piezo transducer was applied in this study to filter the sound generated by the patients during the detection of sleep apnea. Thus the piezo transducer was used at a certain frequency to detect sound at a closer range and serve as the speaker to alert the patient. The experiment was performed in both noisy and quiet closures, under classified investigation.

11.8.3 Jacket

A fabric piece was carefully crafted to suit the objective of this study. Hitherto, other studies have used various means of developing a sleep apnea device, but could not use safer means while using them. However, we used heavy wool-formed materials in the making of the jacket, for easy incorporation of materials.

11.8.4 Bolt IoT and cloud system

The Bolt IoT platform is an online-based module that is connected to the device and sends data in a digitalized form to the Bolt IoT cloud for result interpretation. When a sound is detected by the sound sensor, it smoothly transfers the data to Bolt, and immediately sends the result to a connected phone as a message, and subsequently aggravates the buzzer. In initializing Bolt, codes were created to carry out these processes. Similar to the work performed by the group of researchers on the "project hub" platform, our work predicts and transfers data using the Bolt system.

```
if sensor_value > maximum_limit or sensor_value < minimum_limit:
response = mailer.send_email("Alert", "The Current temperature is beyond the threshold")
if sensor_value > bound [0]:
print ("The temperature increased suddenly. Sending Sms.") response = sms.send_sms ("Someone opened the chamber")
print("This is the response",(response))
elif sensor_value < bound [1]:
print ("The temperature decreased suddenly. Sending an email.")
response = mailer.send_email ("Someone opened the chamber")
print ("This is the response", response)
history_data.append (sensor_value);
except for Exception as e:
print ("Error",e)
time.sleep(10)
```

Furthermore, this device is not only capable of working when connected to the Internet, but it was carefully constructed to suit individuals even without Internet connections.

11.9 Prototype design

11.9.1 Hardware set-up

The components used in the research study include the Arduino micro-controller, ECG sensor, heart sensor, sound sensor, buzzer, Bolt IoT app, fabric, and Bluetooth module. Fig. 11.2 illustrates the organization of materials used in the research study.

11.9.2 Prototype design

After connecting all our components to the jacket as a shield, the device was active in recording analogies in sound at a certain frequency. Thus, when it is connected to an individual's tracheal space, it records the associated rise in sound and sends the results to the respective devices for an alarm. However, limitations were encountered. A trial was conducted on an individual in both sleeping mode and in an act of shouting. This was to enable us to differentiate between the detection of sleep apnea and noise (shouting). There was an immediate buzz from the device in both scenarios. The result equals that of Dhruba and colleagues in their study on IoT-based sleep apnea. Fig. 11.1 shows the image of our prototype work still under construction.

Another study was conducted using polysomnography. The patient has to sleep one night in the laboratory where it is performed. Electrodes in the jacket are placed on the head and the outer edge of the eyelids and the chin and belts are placed around the chest and abdomen. A cannula is placed in the nose to measure airflow, and a sensor is placed on the finger to measure the level of oxygen in the blood. This diagnosis shows eye movement, muscle activity, heart rate, and respiration levels of blood oxygen, airflow, and electrical activity in the brain, which are collected and evaluated.

11.10 Conclusion

Contrary to other studies performed using an electric microphone placed on the tracheal part of the body, our study designed a comfortable sleep apnea detection device that is in the form of a jacket. All the device components were incorporated into a heavy fabric formed as a jacket. Sleep apnea sufferers have a respiratory nightmare characterized by paused and noisy breathing (snoring). This condition is harmful for the body and heart health and requires diagnosis. Usually, people with heart disease are admitted to a sleep disorder center for one night, where various recording devices are connected to their bodies for analysis. The proposed device is large in its current state, we hope to rebuild it in the future to make it smaller and portable. The system is cost-effective compared with the existing systems; hence, it is affordable for everyone.

References

[1] J.G. Webster, E. Halit, Measurement, Instrumentation, and Sensors Handbook: Two-Volume Set, CRC Press, 2018.

[2] J.J. Carr, J.M. Brown, Introduction to Biomedical Equipment Technology, 4th ed., Prentice-Hall, New Jersey, 2001.

[3] I. Sahli, I. Kristanto, T.C. Thali, Perancangan RS 232 to RS 485 converter sistem network multidrop, Jurnal Teknik Elektro 1 (1) (2004).

[4] V. Ravi, V. Udaya, K.V. Bharadwaj, V.B. Adishesha, Simplex high-speed morse coding with ultra low power MSP430, American Journal of Embedded Systems and Applications 7 (1) (2019) 26−40.

[5] S.M. Caples, A.S. Gami, V.K. Somers, Obstructive sleep apnea, Annals of Internal Medicine 142 (3) (2005) 187−197.

[6] H.E. Ransack, S. Redline, E. Shahar, A. Gilpin, A. Newman, R. Walter, et al., Sleep heart health study. diabetes and sleep disturbances: findings from the sleep heart health study, Diabetes Care 26 (3) (2003) 702−709.

[7] P.J. Strollo Jr, R.M. Rogers, Obstructive sleep apnea, New England Journal of Medicine 334 (2) (1996) 99−104.

[8] R.D. Cartwright, Effect of sleep position on sleep apnea severity, Sleep 7 (2) (1984) 110−114.

[9] C. Guilleminault, F.L. Eldridge, F.B. Simmons, W.C. Dement, Sleep apnea in eight children, Pediatrics 58 (1) (1976) 23−30.

[10] J. He, M.H. Kryger, F.J. Zorick, W. Conway, T. Roth, Mortality and apnea index in obstructive sleep apnea: experience in 385 male patients, Chest 94 (1) (1988) 9−14.

[11] A.R. Dhruba, K.N. Alam, M.S. Khan, S. Bourouis, M.M. Khan, Development of an IoT-based sleep apnea monitoring system for healthcare, Computational and Mathematical Methods in Medicine (2021) 7152576.

[12] J.A. Dempsey, S.C. Veasey, B.J. Morgan, C.P. O'Donnell, Pathophysiology of sleep apnea, Physiological Reviews 90 (1) (2010) 47−112.

[13] N.M. Punjabi, The epidemiology of adult obstructive sleep apnea, Proceedings of the American Thoracic Society 5 (2) (2008) 136−143.

[14] K.P. Strohl, S. Redline, Recognition of obstructive sleep apnea, American Journal of Respiratory and Critical Care Medicine 154 (2) (1996) 279−289.

[15] T. Young, J. Skatrud, P.E. Peppard, Risk factors for obstructive sleep apnea in adults, JAMA: the Journal of the American Medical Association 291 (16) (2004) 2013−2016.

CHAPTER TWELVE

Design considerations for diagnostic radiology department

Dilber Uzun Ozsahin[1,2,3], Basil Bartholomew Duwa[3,4], John Bush Idoko[5], Galaya Tirah[6], Nosaiba Elhassan Eldasougi[4], Mohamad Naesa[4], Mubarak Taiwo Mustapha[3,4], Saleem Attili[4] and Ilker Ozsahin[3,7]

[1]Department of Medical Diagnostic Imaging, College of Health Science, University of Sharjah, Sharjah, United Arab Emirates
[2]Research Institute for Medical and Health Sciences, University of Sharjah, Sharjah, United Arab Emirates
[3]Operational Research Center in Healthcare, Near East University, Nicosia/TRNC, Mersin 10, Turkey
[4]Department of Biomedical Engineering, Near East University, Nicosia/TRNC, Mersin 10, Turkey
[5]Applied Artificial Intelligence Research Center, Department of Computer Engineering, Near East University, Nicosia/TRNC, Mersin 10, Turkey
[6]Department of Medical Biology and Genetics, Near East University, Nicosia/TRNC, Mersin 10, Turkey
[7]Department of Radiology, Brain Health Imaging Institute, Weill Cornell Medicine, New York, NY, United States

Contents

12.1 Introduction

Diagnostic radiology is central when considering other methods of diagnosis. This method introduces novel approaches to diagnosing and treating diseases. In Ontario, Canada, an average of eight individuals are diagnosed with life-threatening diseases, such as cancer. Therefore, there is a global deficiency in health professionals, which has affected the diagnosis of diseases and triggered the innovation of sophisticated radiology tools, replacing human efforts.

In the design and layout of radiology and imaging spaces, some important factors have been adopted. The practical requirements for the design depend on the clinical functions of the proposed structure. This study considers opting for different approaches when designing and constructing a standard radiology and imaging room. Therefore radiology has been tremendously active in replacing the human force with tools and machines. It has become necessary to consider the structural approach to designing a supermodel diagnostic radiology room. However, there are some important designs and equipment to consider when constructing a standard diagnostic radiology department [1].

In the last century, the radiology department has experienced the development of numerous new advancements and an expansion in diagnostic imaging units and equipment. For example, resonators and ultrasound imaging are now fundamental pieces of the cutting-edge demonstrative radiology department. The field of diagnostic clinical imaging will continue to advance until it reaches a high-level stage, which will prompt changes in clinical imaging and provide the best treatment to patients. The last 20 years have also seen an increase in the utilization of interventional radiology in a wide range of settings at the second and third degrees of care. Interventional radiology is characterized as the utilization of imaging to direct surgeries, implying that interventional radiology is not utilized for treatment, but as a guide to treatment.

Interventional radiology has replaced some identical medical procedures, as the strategies utilized rely on putting the patient in least danger and utilizing the quickest technique for therapy. Because of the critical expansion in the utilization of computerized innovations, images would now be able to be obtained viably. Strikingly, the entirety of this brought about lower costs. To achieve efficiency in radiological analytic devices, the design of the divisions should be planned with attention, considering the progression of patients inside units or departments. Focus should be given on patient help administrations, for example, conveying patients and waiting areas, to accomplish the potential advantages accessible, which is the goal of the research [2].

12.2 Materials and methods

12.2.1 Modern radiography methods

Diagnostic imaging consists of several imaging methods, which may be general X-ray units, general endoscopy and endoscopy devices, specialized vascular imaging and endoscopic methods, nuclear imaging (Gamma Camera), positron emission photography, X-ray mammography, ultrasound imaging, computed tomography (CT) scan, and magnetic resonance imaging (MRI).

Modern photographic means are characterized by the fact that the images are obtained in digital form, which helps in the archiving and transmission of information, such as picture archives and communications systems (PACS).

12.2.2 Organizing imaging units

1. The department of X-ray and computed tomography should have proper access to the ambulance and emergency department. There should be a close association with outpatient clinics, especially those that involve treating orthopedic patients.
2. A close relationship with pregnancy-related clinics must be taken into consideration during the design phase, especially if the department supports pregnancy diagnostic imaging tests using ultrasound. Emphasis must also be placed on the fact that nuclear radiography and positron emission blocks are not close to the ultrasound suites [3].
3. There is a need for a close connection between the hospital entrance and access to the department, which needs to be directly from the main hospital street on the ground floor of the hospital. This confirms that the outpatient and the general practitioner can easily locate the department. Wide-spaced corridors and doors must be provided between the main entrances and the diagnostic imaging sections to allow the transportation of the equipment, also considering wheelchair accessibility.
4. The availability of the pharmacy to supply analgesic medications is also important. The radioactive materials used in some procedures are supplied by the radiological pharmacy, which is part of a larger pharmacy, or these materials are supplied from a separate part of the imaging department itself.

5. The flank regions should be sufficiently distant from the diagnostic imaging department. Ideally, the design of the movement areas should allow separate access for inpatients who may be on beds or strollers [4].

From the above, we find it is necessary to pay attention to the location of the units in the radiology department and their proximity to other departments. In the next chapter, we explain the design considerations for the diagnostic radiology department.

12.2.2.1 General X-ray unit design
This unit is used for photography in multiple situations, including:
1. Imaging for the diagnosis of cancer and benign and malignant tumors
2. Imaging the chest area
3. Imaging for fractures
4. Arthroscopy to prepare for surgery and postoperative monitoring [4].

12.2.2.2 Patient flow
12.2.2.2.1 Referral
Outpatients are referred by the doctor and attended to according to specific appointments. The patients' data is entered by an employee of the administrative staff (or the radiologist in cases that require a long time). As for inpatients, they may be given priority over outpatients. They are referred from the hospital ward (sometimes after anesthesia) or directly from the emergency department.

12.2.2.2.2 Attending the examination
After confirming the request, it is sent directly to the diagnostic imaging department. It is better to provide separate entries and waiting areas for inpatients. Inpatients still need to be registered with the main reception, and the accompanying staff can assist in the procedures. After the examination is completed, the patient returns to the ward. Patients coming from the ambulance and emergency department are not necessarily in an urgent situation; for these patients, the application form is brought from the ambulance and emergency department to the reception area of the diagnostic imaging department, and they wait in the main or secondary waiting area similar to the outpatient until the completion of the examination. As for emergency and urgent patients, who are often transported by carts, they are examined as soon as possible. These patients are often accompanied by nurses, porters, and, in some cases, doctors. Once the examination is complete, patients are either returned to the emergency

department, sent directly to the ward, or, when required, to the ICU or Cardiac Care Unit. General X-ray rooms used in the accident and emergency (A&E) department should be designed to be larger to allow maneuverability for beds and vehicles. The entrance for patients must be designed to allow entry of beds, carriages, or wheelchairs, with accessories, such as serum holders and other monitoring devices. General X-rays are characterized by a short period compared with other diagnostic imaging procedures, as it can take at least ten minutes to examine each patient [4].

12.2.2.2.3 Special considerations for children
The imaging room must be equipped with adequate radiation protection for the parents or relatives accompanying the patients or who may be present in the room

12.2.2.3 Place elements
The specifications of the adjoining rooms to support the general X-ray rooms should be as follows:
1. The imaging room, which contains the X-ray tube, the patient table, and the vertical board or bucky.
2. An armored control room using fixed lead panels for radiation protection. They can be part of the diagnostic room or in an independent area.
3. A film processing room is not necessary, in some cases, according to the approved technical options.
4. Subwaiting area for both inpatients and outpatients.
5. A toilet for use by patients and their accompanying relatives.
   Other arrangements that may be found in large centers:
1. A porter center to assist in transporting patients to and from the wards.
2. A consulting room.
3. The main reception area for patients.
4. The control room must be within an X-ray room adjacent to the treatment and visualization area for radiation protection and the flexibility of staff movement; the examination room is directly adjacent to it.
5. Patient switching areas and subwaiting areas should be accessible to everyone in the examination rooms. Patient privacy should be given a high priority, and men and women should be isolated from each other through the use of separate facilities.

12.2.3 Description of rooms and equipment

12.2.3.1 X-ray tube

There are two ways of installing an X-ray tube: (1) Attached to a vertical stand and installed on a mobile trolley in the ceiling, which allows the tube to move over a wide range of positions. This combination is preferred due to its inherent flexibility, especially when performing ambulatory images; (2) The X-ray tube can be installed on a floor-standing stand, which can be integrated with the patient's table and the X-ray generator. This method saves space but is less flexible. This unit is used when there is limited space. It is, however, not recommended in the ambulance and emergency department, which requires more flexibility [4].

12.2.3.2 Patients' table

The diagnostic image is obtained in most examinations with the patient lying on the table and lying down according to the imaging area of their body. The equipment installed on the ground is placed in the middle to facilitate the entry and transport of the patient from the vehicles.

12.2.3.3 Chest imaging bucky

It is used to acquire images of the chest and take pictures of brutal exposures of a patient standing or lying down. They are installed in most X-ray rooms. The chest plate is usually located near the wall of the examination room and has a holder attached either to the ceiling or to the floor. The cassette holder dimensions are approximately 200 mm x 750 x 800 mm and can encompass a conventional or digital X-ray cassette with dimensions of 35 x 43 cm. Some buckets require electrical feed for the vibrating grid and AEC devices integrated with the quantum system. Some of these units are installed in such a way as to allow the rotation of the cassette or bucky holders at a 90-degree angle and the movement in the horizontal direction to examine the extremities, they must provide sufficient space to allow the movement of these devices and expose the patient to radiation either standing or lying down, and it is recommended to leave a distance of at least 2 m between the base of the patient's table and the surface of the chest bucky.

12.2.3.4 X-ray generator and documentation rooms

Usually two rooms are installed within the X-ray room. In the first room is the kilovolt and high-frequency generator equipment used to operate the X-ray tube, while in the other room, the equipment manuals are stored. For some installations, the beam generator can be installed under the X-ray table and this is ideal in limited spaces. Emergency shutdown switches should be located close to the patient table in the examination room and easily accessible by the operator in the control room [1].

12.2.3.5 Control area

In practice, X-ray dose settings, start-up settings, and other parameters are entered using the control room. This room is usually placed 5.0/5.0 opaque/lead radiation-shielded, forming a separate control area within the X-ray examination room. The radiation protection window in its lower section should have a solid structure with a height of 1.1 m at most, and the upper part should be made of leaded glass that protects against radiation, which allows an adequate view of the patient at all times. The total height of the window should not be less than 2 m from the end of the floor surface. The window shall be installed to allow floods from the examination room and does not include the interior doors. Usually, general X-ray rooms are shot with a thickness of mm^2 per 150 kV. The control room includes wall-mounted electrical distribution panels, main off/on switches, and emergency stop presses. There may be two radiologists in the control area and other people may accompany them.

12.2.3.6 Switching area

Patient switching facilities should be provided close to the radiology room, and the project team must take into consideration the following points for switching rooms in the general radiology department:
1. Individuals switching rooms are adjacent to the examination room and open directly to them.
2. Changing rooms are grouped, but they are not adjacent to the examination rooms, but rather within the subwaiting room where patients can wait or change their clothes before being admitted to the X-ray room.
3. One of the chambers should be designed to allow the entry of a C-Wheel and aid in the exchange process.

12.2.3.7 Secondary waiting areas

Three seats must be provided for each changing room within the secondary waiting area, and if baskets are used to place items, the number of seats increases to five.

12.2.3.8 Modes of providing X-ray services

The modes of providing X-ray services to support a wide range of ambulance and emergency departments are as follows:

1. General X-ray examination rooms used in ambulances and emergency department, need to carry out necessary treatment. The exposure area should be separate from the diagnostic imaging department. The rooms are similar to the general X-ray rooms, but they may differ and have easy and rapid movement of patients on vehicles or beds within the examination room.

2. Alternatively, the diagnostic imaging unit of the ambulance and emergency departments is combined with the diagnostic imaging department, but also has direct access to the ambulance and emergency department. In this case, plans need to be put in place for the emergency department and the diagnostic imaging department to be adjacent to each other. Within the broader diagnostic imaging department, general X-ray and CT scan rooms are as close as possible to the A&E department, taking into account other design considerations, such as the relationship with the fracture clinic. Fig. 12.1 demonstrates the proposed design.

12.2.4 X-ray mammography

A mammogram machine is a device for mammography by X-ray. It is different from a traditional X-ray machine because the radiation energy in a mammography machine is less than that of a traditional X-ray machine. A mammogram device is used in the routine medical examination of breast cancer for women in the age groups at risk of developing cancer (without the appearance of symptoms for women). It is also used to photograph women with symptoms that indicate the possibility of a lump in the breast [5].

Mammography is a sensitive technique used as a primary diagnostic method for breast cancer. Although this technique is very sensitive, it is not relied upon to determine the type of lesion, and for this purpose, a biopsy and analysis are performed. Ultrasound and MRI are used as complementary methods for mammography. These imaging centers are located within the breast care center or the diagnostic and interventional imaging facilities in the radiology department. Because mammography devices are compact and smaller than traditional X-ray machines, they require a much smaller space, and there is also a great need for patient privacy, which should be taken care of in the design of the room.

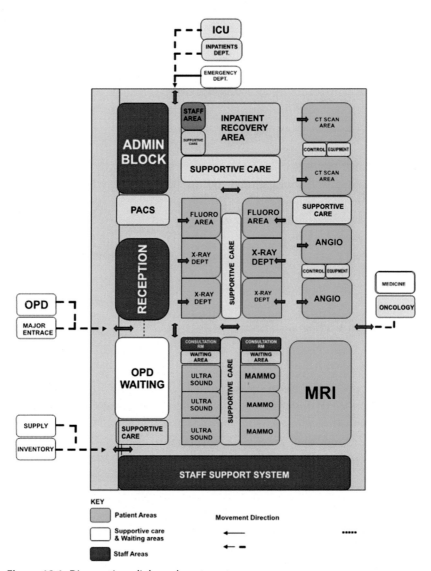

Figure 12.1 Diagnostic radiology department.

12.2.4.1 Early diagnosis of breast cancer

The early screening program for breast cancer is an attempt to achieve an early diagnosis of breast disorders in women between the ages of 50 and 64 years old. The mammography technique was chosen for this purpose after extensive reviews of the available information from alternative imaging systems. X-ray is a special challenge in terms of designing facilities as it involves reading films and preparing reports. Radiologists and others trained in reading X-ray films receive a large number of mammogram images daily, and therefore the image reader must have high skills and a strong ability to focus to obtain the most accurate results. The environment in the film reading and reporting facilities is designed to maximize the visual acuity and characteristics of the conditions. Other things, such as external sounds or interruptions by colleagues, should be avoided [5].

12.2.4.2 Place elements for X-ray mammography

1. Mammography rooms should be proportional to the number of patients expected to attend
2. Reporting rooms for radiologists
3. 3D ultrasound room
4. A histologist's room, which serves more than one mammogram room in which the biopsy is performed
5. Private waiting room
6. Chalets
7. Subpatient waiting areas

12.2.5 Ultrasound

Ultrasound imaging is based on the principle of nonionizing unlike ionizing radiation, and therefore ultrasound imaging is less dangerous than other imaging procedures, including MRI (nonionizing scans), for both the operator and the patient, as it operates at frequencies between 5.3 MHz and 20 MHz. The choice of location for the ultrasound section depends on the need to be far from the magnetic field fringes of any MRI machine, as fields as low as 5.0 gauss can interfere with the operation of the imaging probes. If it is suggested to place the wing near the resonator devices, including those floors above and below the devices, then an evaluation of the magnetic field fringes should be made [4].

Ultrasound units are used in many fields, including:

12.2.6 Computed tomography scan

A CT scan is a form of tomography that combines X-ray images from several projections into single or multiple images. Instead of using an X-ray beam receiving film, a row of solid reagents is used to extract data from different projections. The doses in some operations can be relatively high when compared with other diagnostic imaging techniques, and thus there is a need to block the X-rays in the examination room. It is possible to place the CT suite in combination with the MRI unit, thus achieving volume savings, especially in support areas.

12.2.7 Magnetic resonance imaging devices

MRI is mainly performed by placing the entire body of the patient or a specific part of the body within a magnetic field of high intensity, ranging in modern applications between 1.0 and 4 Tesla [6]. These devices are used to obtain moving images to examine the anatomical and physiological structure. The main difference between comprehensive imaging equipment and telescopic systems lies in the method of operation. The comprehensive endoscopy equipment is operated by the radiographer or radiologist using control devices. All movements of the device are controlled together in addition to exposure of the endoscopy, and should not be present in the examination room when the X-ray tube is issued [4].

12.2.8 Contrast materials

Modern radiography uses contrast materials in all imaging methods, including MRI, CT, and ultrasound scans. Contrast materials are used to improve and increase the contrast of the images obtained in personal imaging scans.

12.2.8.1 Examination room

The image light intensity provider and a ray tube (X) are installed on a fixed (U) letter arm that is connected to the top and bottom of the patient's table in some examinations, such as side or recumbent exposures. When the patient lies on the table during barium enema examinations, the X-ray tube and image light intensity supplied are integrated with the unit. It does not support the required radiation projection, so there must be an X-ray tube attached to the ceiling, and it is preferable to have at most two screens within the X-ray examination rooms and one screen within the control rooms. In most cases, there must be one generator in the room to provide power to each of the two X-ray tubes present. The space should allow for four cabinets as a whole, one for brochures, one for endoscopy and digital raster imaging, and two for the generator and power distribution unit [7].

12.2.8.2 Control room

The control room can be in the examination room or in a separate room. It has the same layout and structure as the general X-ray suite, but it needs to be larger to include more control and monitoring equipment. In addition, the following points must be taken into consideration [7]:

1. The presence of a monitor screen with the imaging computer and the user interface.
2. A video recorder to store the blank videotapes.
3. Space must be provided for the presence of four intermittent medical staff.

12.2.8.3 Barium disposal

Unused barium cannot be disposed of by traditional laundries and sinks, so a private washroom must be provided within a dedicated or shared unit for dirty supplies [7].

12.2.8.4 Endoscopic equipment for angiography and nonvascular imaging and interventional procedures

This type of endoscopy equipment is suitable for hospitals in areas of general specialization that do not contain cardiac imaging facilities or other specialized imaging facilities. This device is used for angiography, avascular imaging, and interventional procedures, and angioplasty and endoscopy are used in several areas [8].

12.2.8.5 Patient flow

A room, office, or any other facility must be provided for discussion and obtaining the consent of the patient, and it must be either within the department or the interventional ward. The patient under anesthesia needs blood pressure, oxygen saturation, and cardiogram monitoring during the interventional procedure. This can be done by using a monitor installed on a cart, unlike diagnostic imaging. Earlier the interventional procedures took a lot of time and the duration of the operation could not always be predicted. It may be a short procedure of half an hour, or it may extend up to three hours. The average duration of an hour or an hour and a half is appropriate after the examination. The patient can be transferred from a radiation table to a cart or bed and then to the recovery room [8].

12.2.8.6 Examination room

There is no need to have an X-ray tube attached to the ceiling with this type of equipment. The rooms include a set of integrated display screens of up to four screens, and they must move throughout the field and be installed on a roof trolley, the same one used to support radiation tubes X installed on the ceiling. There is a need for a special clinical operation light for interventional procedures. During some procedures, particularly those involved in general anesthesia, up to nine employees may be present in the examination room and control area.

12.2.8.7 Control room

Most procedures are performed with the radiographer, radiologist, and other medical staff in the main procedure, and it is still necessary to have a separate and adjacent control room to:

1. Installing a separate unit for radiation control unit X to operate the system if the control unit in the examination room fails.
2. Show the obtained images after the end of the procedure or during it, and after that, digital or paper copies can be made.
3. The presence of doctors and students to monitor the procedure.

The designer should consider that there are two options for designing the control room associated with the interventional rooms:

1. The control room can be designed similarly to radiation room X by providing radiation protection panels.
2. A separate room can be provided for the installation of control equipment and the imaging station. Direct access and direct monitoring must be provided in the examination room. This area can be used as a control room for more than one imaging and endoscopy room.

12.2.8.8 Common area for anesthesia and postoperative recovery

These areas should be provided at a rate of two rooms per examination room for patients who need anesthesia or resuscitation after anesthesia and can also be used to administer medicines to patients before starting the procedure and to administer sedatives. It should also be designed to maintain patient privacy.

12.2.9 Nuclear medicine imaging facilities

1. Radioactive materials are used in gamma camera imaging and nuclear medicine, where a pharmaceutical compound is tagged with radioactive isotopes and then given to the patient. After a certain period,

images that show the distribution of radioactive materials inside the body are taken by the gamma camera device.

2. Gamma imaging uses a camera to examine a wide range of organs of the human body (imaged) with a gamma camera, such as the brain, kidney, lungs, heart, and bones, as well as to detect tumor spots in the tissues. Therefore, the gamma imaging section should be integrated with the department of cancer, heart diseases, and respiratory diseases.

3. Most of the patients who receive images are over the age of 50, so these imaging centers must have enough space to place resuscitation devices in areas where the risk of radiation and pollution is low, except for imaging centers for children [4].

12.2.9.1 Location of gamma camera facilities

It is beneficial for the gamma imaging department to place a camera near or within the central radiological diagnostic (imaging) department, but in some cases, the floor loads and radiation protection requirements play the largest role in determining the location, and this should be carefully researched with the radiologists, medical physicists, and appropriately qualified consultants or radiation workers in a gamma–imaging facility. To reduce the peril of radiation exposure, you should avoid approaching the imaging section of the gamma camera from the following sections:

1. Dental X-ray rooms and other imaging methods, such as MRI rooms, attended by a significant number of children.

2. Echo ultrasound room and obstetrics departments.

The consultation rooms, offices, reception, waiting areas, and nonradioactive materials store should be considered depending on the number of nuclear medical examination rooms. These facilities can also be shared with other diagnostic techniques within the department [4].

12.2.9.2 Radiators pharmacy

The production of radioactive pharmaceutical compounds requires highly specialized facilities, in addition to trained and qualified staff for this purpose. These radioactive pharmaceutical compounds are produced in one or two local centers that provide them to other hospitals to carry out the gamma imaging process. The radiant pharmacy can be a part of the general pharmacy However, this will increase the radiation protection requirements for workers and the facility, and the relationship between the imaging department and the radiators pharmacy will become critical.

12.2.9.3 Positron emission tomography units without a rotational scrambled particle accelerator

Positron emission tomography (PET) diagnostics provide clinical and physiological information about the imaged organs. Positron emission imaging is similar in principle to gamma imaging with a camera. However, the gamma-ray energy emitted in the positron emission is four times greater, so there are additional requirements for radiation protection when building new facilities that must be taken into account when designing the gravity of handling these radioactive materials:

1. The production of radioisotopes used in the positron emission requires a rotational accelerator (cyclotron), but the half-life of these isotopes is short (1−2) hours, so the facilities for the positron emission must be close to the rotational accelerator and can be reached quickly.
2. The substances most often used are fluorine-labeled glucose-18. These radioisotopes and others are used in positron emission to detect clinical ailments that are difficult to detect by other techniques.

12.2.9.4 Positron emission imaging rooms

1. The positron emission imaging room is located on the ground floor due to high loads. This room is similar in design to a normal gamma camera room in terms of equipment and connection cables.
2. A medical washroom should also be available, taking into account that it is not for the disposal of liquid radioactive waste.
3. Space must be provided to place cabinets for storing positron emission calibration equipment, and they must be armored with bullets of 5 mm thickness to avoid radiation hazards.

12.2.9.5 Control room

1. It is adjacent to the X-ray scanner room and linked to a monitoring window and a door. It is used to manage the imaging process, monitor the patient during imaging, control the gamma camera, and perform the initial review of the acquired images.
2. Space must be allocated to store acquired images, either in digital or printed form.
3. Shelves should be designated to hold papers and brochures.
4. Air conditioning must be provided in the control room.

12.2.9.6 Injection chamber

1. The design must be taken into account so that patients in wheelchairs are allowed to enter the injection room, and the design must also be comfortable and relaxing for the patient.
2. Armored areas must also be available for storing radioactive materials or previously prepared radiation doses. Regarding protection from radiation, they are similar to those in the positron emission imaging rooms (in terms of walls, ceiling, floor, and finishes).

12.2.9.7 Patients' toilets

It should have proper access for disabled people. It has the same design as the Gamma Camera Suite.

12.2.9.8 Radioactive material preparation area

1. The materials and finishes in this room are nonabsorbent and easy to clean.
2. The radiation monitoring equipment is kept to monitor radioactivity.
3. It is not recommended to design windows in the preparation area.
4. There should be a disposal drain for radioactive liquid waste, and this drain should be similar to the radiator's pharmacy in the Gamma Camera Suite.
5. Warning signs should be placed at the entrance of the preparation room to warn of radiation hazards [9].

12.2.9.9 Engineering requirements

This appendix describes the engineering services present in the imaging facilities and how these services are integrated with the engineering systems to fully serve the site, as the focus is on the proper installation of the equipment within an appropriate environment to achieve high standards of performance and reliability in imaging.

12.2.9.10 Environmental requirements

The appropriate environmental requirements for the work of devices and equipment must be obtained in detail from the suppliers, as the comfort of the patients and the work team is the most important consideration regarding ensuring the stability of the temperature and the effect of the heat derived from the diagnostic imaging systems. The ability to control temperature and humidity is the basic factor in the success of the design. Air conditioning should be preferably self-contained domestic units [10].

12.2.9.11 Space allocated for generators, transformers, computers, and machines

Suitable spaces must be allocated to place electronic equipment such as generators and equipment cabinets, and these spaces must be designed in a way that supports easy access to the equipment and is kept far from the hands of unauthorized persons.

12.2.10 Mechanical services

12.2.10.1 Heating

1. Heaters should be used in places that rely on low-pressure hot water systems for heating.
2. Hot water pipes exposed to contact should be insulated.
3. Radiators should be placed under windows or on exposed walls, with an appropriate distance being taken between the top of the radiator and the edge of the window to prevent the heat produced from the radiator from being reduced by the curtains. A sufficient distance must be left under it to allow the use of cleaning tools.
4. It is recommended that the radiators be proportional to the valves of the thermostats, which must be of a sturdy structure and selected in a manner that matches the pressure and temperature specifications of the heating system serving the various sections.

12.2.10.2 Ventilation

1. Some areas need mechanical ventilation for clinical or functional reasons, regardless of their location, whether internal or peripheral.
2. Those in temporary accommodations need less or no mechanical ventilation.
3. Most places within the imaging department need mechanical ventilation due to the heat from the equipment, the preparation of patients and staff, and clinical reasons.
4. Flow ports must be defined to obtain a uniform distribution of air within a vacuum without disturbing patients.
5. Air suction and supply systems must have indicator lights to confirm the operating status of each.

12.2.10.3 Medical gas tubes

It should be provided with rooms that may require the use of nitrous oxide or other gases.

12.3 Conclusion

In the clinical setting, one of the most important units is the diagnostic radiology unit. In this paper, we have designed a perfect and error-free diagnostic radiology department. In our preparatory investigation, we carried out a detailed study of the diagnostic and interventional radiology units, considering the mechanical, environmental, and electrical engineering requirements, following the British standard. The role of the medical engineers in studying the effect of device placement on the design and the effect of design on the placement of devices in addition to the flow of patients within the department have been studied. The proposed design, if applied to hospitals across the globe, would reduce treatment turnaround time and increase patients' confidence when visiting the radiology department.

References

[1] A.R. Padhani, L. Ollivier, The RECIST criteria: implications for diagnostic radiologists, The British journal of radiology 74 (887) (2001) 983–986.
[2] R.F. Mould, The early history of x-ray diagnosis with emphasis on the contributions of physics 1895–1915, Physics in Medicine & Biology 40 (11) (1995) 1741.
[3] K. Doi, Diagnostic Imaging Over the Last 50 Years: Research and Development in Medical Imaging Science and Technology, Institute of Physics Publishing, 2006.
[4] NHS Estates, "Facilities for Diagnostic Imaging and Interventional Radiology," vol. 1, London: The Stationery Office, 2001 pp. 252.
[5] K.F. Defreitas, B. Ren, C. Ruth, I. Shaw, A.P. Smith, & J.A. Stein, U.S. Patent No. 7, 831, 296, Washington, DC: U.S. Patent and Trademark Office, 2010.
[6] J. Karpowicz, K. Gryz, Health risk assessment of occupational exposure to a magnetic field from magnetic resonance imaging devices, International Journal of Occupational Safety and Ergonomics 12 (2) (2006) 155–167.
[7] C.L. Smith, U.S. Patent No. 3,100,870, Washington, DC: U.S. Patent and Trademark Office, 1963.
[8] D.J. Spinose, J.A. Kaufmann, G.D. Hartwell, Gadolinium chelates in angiography and interventional radiology: a useful alternative to iodinated contrast media for angiography, Radiology 223 (2) (2002) 319–325.
[9] P.D. Olcott, A Positron Emission Tomography (PET) Neuro Insert for Combined PET/Magnetic Resonance Imaging, Stanford University, 2014.
[10] Y. Zhang, S. Chen, Y. Cai, L. Lu, D. Fan, J. Shi, et al., Novel x-ray and optical diagnostics for studying energetic materials: a review. *Engineering* 6 (9) (2020) 992–1005.

CHAPTER THIRTEEN

Construction of an automated hand sanitizer dispenser used against transmissible diseases

Dilber Uzun Ozsahin[1,2,3], Basil Bartholomew Duwa[3,4],
Declan Ikechukwu Emegano[3,4], Mubarak Taiwo Mustapha[3,4],
Natacha Usanase[3,4], Efe Precious Onakpojeruo[3,4] and
Ilker Ozsahin[3,5]

[1]Department of Medical Diagnostic Imaging, College of Health Science, University of Sharjah, Sharjah, United Arab Emirates
[2]Research Institute for Medical and Health Sciences, University of Sharjah, Sharjah, United Arab Emirates
[3]Operational Research Centre in Healthcare, Near East University, Nicosia/TRNC, Mersin 10, Turkey
[4]Department of Biomedical Engineering, Near East University, Nicosia/TRNC, Mersin 10, Turkey
[5]Department of Radiology, Brain Health Imaging Institute, Weill Cornell Medicine, New York, NY, United States

Contents

Practical Design and Applications of Medical Devices.
DOI: https://doi.org/10.1016/B978-0-443-14133-1.00023-9

13.1 Introduction

The Centers for Disease Control and Prevention (CDC) suggest washing hands as often as possible with soap and water for about 20 seconds, especially after being in public places or after blowing your nose, coughing, or sneezing. When soap and water are not readily available, the CDC is of the opinion that an alcohol-based hand sanitizer of 60% is recommended. As the coronavirus spread over the world, demand for hand sanitizers skyrocketed. Gel-like hand sanitizer is applied by pressing a pump with one's hand [1]. The world is currently experiencing a pandemic caused by coronavirus illness (COVID-19), which was first found in December 2019 in Wuhan, China. However, the World Health Organization (WHO) termed the virus as very contagious and recommended some advice to help limit community transmission in a variety of ways. Hand washing with soap or hand sanitizer on a regular basis is one of the obligatory suggested measures [2].

Good hygiene involves regular hand washing at home and in our different communities, which helps to reduce the occurrence of infections, especially those that invade the intestines (gastrointestinal), respiratory tract, and skin. Reducing the bacteria load in our hands can be accomplished with soap and water or by using hand sanitizers free of water, which minimize contamination. This removes or kills the organism in question. The use of products and practices that help to reduce both bacteria and virus loads on hands, singly or in sequence, augments the health impact of hand hygiene within a particular community. This could be achieved by persuading people to practice hand cleanliness, thereby increasing hand hygiene [3].

Washing hands regularly is now widely accepted as one of the most essential aspects of infection prevention and control. Among the surging severity of diseases and treatment complexity, as well as the pandemic faced globally, which is fueled by multidrug-resistant pathogens, healthcare professionals strive effortlessly to revert back to the basic ideas of preventing infection by practicing hand hygiene [4].

13.2 Related work

13.2.1 History of hand sanitizer

According to some reports, the first-hand sanitizer was invented by Lupe Hernandez, a Bakersfield nursing student, in 1966. Even without soap or water, she discovered that a 60%−65% alcohol solution can destroy some bacteria and viruses. Her invention of hand sanitizer dates back to 1966. Reports have it that she devised sanitizer because medical professionals lack soap and water. Despite her efforts, some authors believe it was a myth because it lacked strong evidence that she invented the product because her name was not even on a hand sanitizer patent in the United States [5].

Hand sanitizer is composed of isopropanol, a mixture of isomers, such as ethanol-propanol, which is the most common active component in alcohol-based hand antiseptics. The property of alcohol that makes it antibacterial is its ability to denature and coagulate proteins. These organisms lose their protective covering and become inactive as a result. The CDC recommends sanitizers with an alcohol content of 80% (percent per volume) ethanol or 75% isopropyl alcohol; however, hand sanitizers with an alcohol content of 60% to 95% are generally suitable for decontamination [6].

In particular, larger than recommended amounts are paradoxically less powerful since proteins do not quickly denature when water is present. Hand sanitizer made with alcohol are usually given as a percentage of volume, rather than a percentage of weight. A 15 seconds contact duration was able to reduce bacteria by more than 5 log10 steps in a study using 85% (weight/weight) ethanol [7]. Ethanol, which is the most common constituent in alcohol, seems to be the most efficient against viruses, whereas its isomer (propanol) is thought to be a better bactericidal. Alcohols, when mixed together, may have a synergistic impact on organisms.

The effectiveness of hand sanitizers is dependent on the amount of alcohol present. A study shows that a hand rub often contains 85% ethanol and is much more effective in reducing bacterial populations than treatments containing 60% to 62% ethanol. Behavioral Health Services (BHS) frequently include humectants, such as glycerin, which assist in avoiding skin dryness, as well as moisturizers, such as aloe vera, which aid in replenishing the water lost during use. The alcohols mentioned do not cause acquired bacterial resistance; hence they are regarded as extremely effective for usage in medical contexts [8].

According to Matatiele et al. [9], alcohol-based hand sanitizer (ABHS) is a beneficial item for preventing the transmission of infectious viruses in crowded environments, such as clinics, workplaces, and schools. It also aids in the prevention of disease-causing organisms (bacteria) from spreading. ABHS considerably decreases bacteria counts on hands, according to early extensive studies on the efficiency of antiseptic hand rubs [10]. According to the study, ABHS is more effective than a plain soap hand wash at preventing gram-negative bacteria transmission. The hand sanitizer dispenser is important because it allows people to wash or rub their hands with ABHS while they are moving. According to Arnab Das et al. [6] when a hand sanitizer is strategically placed, its effectiveness increases hand cleanliness activity from 1.52% to above 60%. To deliver liquid or gaseous sanitizing agents, a variety of dispensers are available, including mechanical, automated with pushbuttons, touchless, and so on. Mechanical dispensers are commonly used in public areas, including hospitals. Because mechanical dispensers require physical contact, they are susceptible to pathogen contamination. By conducting research on a mechanical hand sanitizer dispenser used in hospitals, researchers determined that when a sick person touches the dispenser, it could be contaminated, leading to hospital-acquired infections [11].

Automated hand sanitizer dispensers are used in healthcare settings; however, they can become polluted and serve as a breeding ground for infections. According to studies, previous mechanical and electrical dispensers (with pushbuttons) are at increased risk as they can be contaminated with germs that cause hand-related diseases. As a result, automated touchless sanitizer is now being used in hospitals, particularly in industrialized countries. Because this dispenser operates without human interaction, it has the potential to be very effective in preventing the transmission of infectious diseases if used properly. Since several types of sensors detect the closeness of objects, a sanitizer dispenser could be touchless and automatic in many ways [12]. As a result, automated touchless sanitizers are now being used in hospitals, particularly in industrialized countries. The dispenser lacks the human experience to use it, so it has the potential to be very effective in preventing the transmission of infectious diseases if used correctly. A sanitizer dispenser can be constructed to be touchless and automatic in a variety of ways, as numerous sorts of sensors can be used to measure proximity. In general, ultrasound sensors along with infrared are utilized in making a sanitizer dispenser that is cheap. However, they function poorly in public because of noise interference. Some dispensers employ sensors made with infrared radiation, although it has adverse effects, especially during sunny days when the strength of sunlight varies due to clouds [13].

However, by utilizing a resistor that is light-dependent or a photo-resistive sensor, such limitations can be easily solved. A laser light is utilized in this study to prevent other light reflections in the sensor that resists light, and this application of laser-based closeness tracking with a light-dependent resistor yielded more successful and user-friendly results when employed in crowded public locations. With the number of people living in low-income countries, several public locations, such as bus stops, train stations, raw markets, and hospitals, are frequently packed. The general public either lacks the skill or is indifferent about maintaining individual sanitization in a clean environment [14].

The primary goal of this research is to make it easier to assemble and manufacture a hand sanitizer dispenser that is low cost, free with touches, and automated utilizing laser detection technology.

This study has a unique way of designing a dispenser that is automatic, cheap, and readily available in every developing country. Inside the laser chamber, the photoelectric resistor detects human hands, and this sensor works well in both daytime and darkness.

To demonstrate the benefits and drawbacks of the manufactured device, a full comparison of it with traditional devices is offered. As a result, this research could aid in the prevention of Covid-19 transmission in highly populated poor countries. Some related studies by Sathwara et al. adopted the use of liquid-crystal display (LCD) in the development of the sanitizer dispenser. The LCD has a flat-panel display that manipulates light. The first mass-produced LCD panel technology is known as TN (Twisted Nematic). The theory underlying the LCD is that when the liquid crystal molecules in the LCD cell are not exposed to an electrical field, the molecules twist 90 degrees. The light is polarized and twisted with the liquid crystal molecular layer when it goes through the first polarizer, whether from ambient light or from the backlight [15]. It is blocked when it reaches the second polarizer. The display seems black to the spectator. The liquid crystal molecules are untwisted when an electric field is introduced to them. When polarized light reaches the layer of liquid crystal molecules, it is not bent and travels straight through. When it goes into the polarizer, the liquid crystal molecular layer twists and polarizes it. It will likewise travel through the second polarizer, and the spectator will notice that the display is bright [16]. The ultrasonic sensor device utilizes ultrasonic sound waves to determine the distance of objects. It has a transducer that receives impulses as messages about the distance of an object. Some sound waves with high frequencies reverberate, causing echoes [17].

13.3 Materials and methods

13.3.1 System architecture

The research was conducted in an enclosed biomedical laboratory and a 3D bioprinting laboratory, where we designed our model bottle handler. The components were collected based on the aim of the research in the online stores. Fig. 13.1 shows the schematic representation and architecture of the method adopted in this study. The components are connected to one another for the proper functioning of the system.

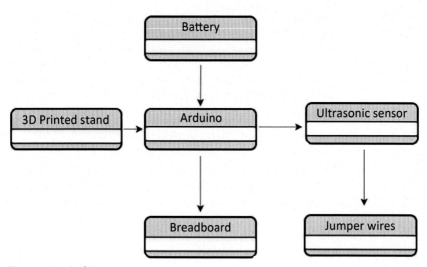

Figure 13.1 Architecture.

13.3.2 Components

The components were purchased from various online stores, such as Amazon and Digi-Key Electronics. The prices were weighed and compared for each component in different stores before purchase, as illustrated in Table 13.1.

 13.4 Results

13.4.1 Ultrasound sensor

Ultrasonic sensors (previously known as OsiSense) detect targets in a contactless, dependable, and high-performance manner, regardless of the shape, texture, color, or transparency of the target. The article further reiterates that its operation is straightforward. The ray emitted by the ultrasonic pulse is about 40 khz, and this travels through a vacuum and refracts when it encounters an object. The sensors are optimal for

Table 13.1 Components and their functions.

Components	Functions	Amount of purchase ($)
Arduino UNO microcontroller	Essential in pressing the data from other components to process it in pumping the bottle of liquid.	23
Ultrasonic sensor SR04 or distance sensor	Detects the distance between the persons approaching the device and prompts for sanitizer delivery.	3.95
3D-printed container handler	A suitable container handler that contains the sanitizer	Approximately 3
Jumper wire 28 AWG wire	Essential in connecting the Arduino UNO with other components of the IoT sensors.	2.25
AC adapter 7.4 V lithium battery balance charger	Important in generating electric current for the device to work efficiently.	12
Breadboard	Connects all the components for a flawless electric flow.	8.65
Accessory screw	Helps in knotting the components together to the 3D bottle handler.	2

Figure 13.2 İmage of ultrasonic sensor.

Figure 13.3 Arduino Uno microcontroller.

detecting clear objects. This works irrespective of the color or texture of the object, as can be seen in Fig. 13.2.

13.4.2 Arduino Uno microcontroller

The Arduino Uno microcontroller system utilized in this study is connected to the various components of the proposed dispenser. As shown in Fig. 13.3, the device is known to be efficient and cheap, with 14 digital pins and 6 analog pins.

13.4.3 3D model

The 3D-designed model of our bottle holder was designed using the Creality 3D CR-10 printer, which is one of the most preferred and affordable 3D machines, particularly for this kind of study. We

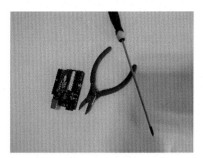

Figure 13.4 Accessories and screws.

downloaded the design match in a 3D online platform called MakerBot Thingiverse for our preferred design.

13.4.4 Accessories and screws

We adopted the application of screws for knotting our components together. This has enabled us to have organized work, as shown in Fig. 13.4.

13.4.5 Jumper wire

This component is important for connecting our various components together for an efficient flow of current around the channels.

13.4.6 Breadboard

The breadboard coordinates the various components for an efficient and organized movement of data for a successful outcome.

13.4.7 Prototype

The research objective was achieved through the design and construction of a suitable, affordable, and efficient touchless device model used in the dispensing of sanitizer. The device, as seen in Fig. 13.2, demonstrates similar attributes to that of the Luxton automatic dispenser and Purell sanitizer dispenser found on Amazon's online market, which is more expensive than our designed device. The challenges we faced were the delay in the supply of our study materials that we ordered online and the inefficiency of the device when it was first operated. As shown in Fig. 13.5, our prototype illustrates a highly modern device that is considerably cheaper and more affordable than the ones used.

Figure 13.5 The Prototype.

13.4.8 Program

```
int trigger pin = 4
int echo pin = 3
long distance;
void set up (1)
(
pinMode (trigger pin, OUTPUT);
pinMode (echo pin, INPUT);
serial.begin (9600);
)
Void loop ()
)
Digital Write (trigger Pin, Low);
delay Microseconds (2)
\\clearing the trigger
digital Write (trigger pin, HIGH);
delay Write (trigger pin, Low);
\\capturing the time distance for sound wave to travel microseconds
duration #pulseIn (echo pin, HIGH);
Serial.printIn (distance)#
```

As presented in the program of our prototype, the simulation command was based on the echo and trigger sensors, the object is sensed at about 2 microseconds at 4 cm from the dispenser. The echo sensor sends the signal to the microcontroller, activating the pump to release the liquid using the code based on the distance attained. This simulation was guided by the Maker.io [18] online platform, which made a prototype dispensing machine using a servo motor.

13.5 Conclusion

Disinfecting gel is without doubt a major element that assists in controlling germs and bacteria. Thus, with these effects, we understand their tremendous application in controlling the spread of diseases. A report by CDC elaborated on the importance of cleaning hands and surfaces with hand sanitizers instead of soaps, "Soap and water act in removing germs, while the sanitizers work in completely destroying the germs." Similarly, due to the spread of diseases, advanced methods have been devised to contain them using novel technologies, such as developing a self-detecting device to dispense the substance to the individual at a certain distance. We understand that the existing machines (dispensers) were expensive, hence the reason we focused our research on developing more affordable and easier-to-handle dispensers.

References

[1] R.L. Jess, C.L. Dozier, Increasing handwashing in young children: a brief review, Journal of Applied Behavior Analysis 53 (3) (2020) 1219–1224.

[2] J. Riou, C.L. Althaus, Pattern of early human-to-human transmission of wuhan 2019 novel coronavirus (2019-nCoV), December 2019 to January 2020, Eurosurveillance 25 (4) (2020) 2000058.

[3] L.D. Moore, G. Robbins, J. Quinn, J.W. Arbogast, The impact of COVID-19 pandemic on hand hygiene performance in hospitals, American journal of infection control 49 (1) (2021) 30–33.

[4] A. Broncano-Lavado, G. Santamaría-Corral, J. Esteban, M. García-Quintanilla, Advances in bacteriophage therapy against relevant multidrug-resistant pathogens, Antibiotics 10 (6) (2021) 672.

[5] R. Sequinel, G.F. Lenz, F.J.L.B.D. Silva, F.R.D. Silva, Alcohol-based solutions for sanitizing hands and surfaces in the prevention of covid-19: informative compendium from the point of view of the chemistry involved, Nova Chemistry 43 (2020) 679–684.

[6] A. Das, A. Barua, M. Mohimin, J. Abedin, M.U. Khandaker, K.S. Al-Mugren, Development of a novel design and subsequent fabrication of an automated touchless hand sanitizer dispenser to reduce the spread of contagious diseasesIn Healthcare 9 (4) (2021) 445. Multidisciplinary Digital Publishing Institute.

[7] J. Zhao, B. Zhou, J.P. Butler, R.G. Bock, J.P. Portelli, S.G. Bilén, IoT-based sanitizer station network: a facilities management case study on monitoring hand sanitizer dispenser usage, Smart Cities 4 (3) (2021) 979–994.

[8] R.Y. Booq, A.A. Alshehri, F.A. Almughem, N.M. Zaidan, W.S. Aburayan, A.A. Bakr, et al., Formulation and evaluation of alcohol-free hand sanitizer gels to prevent the spread of infections during pandemics, International Journal of Environmental Research and Public Health 18 (12) (2021) 6252.

[9] P. Matatiele, B. Southon, B. Dabula, T. Marageni, P. Poongavanum, B. Kgarebe, Assessment of quality of alcohol-based hand sanitizers used in Johannesburg area during the CoViD-19 pandemic, Scientific Reports 12 (1) (2022) 1−7.

[10] I. d'Angelo, R. Provenzano, E. Florio, C. Pagliuca, G. Mantova, E. Scaglione, et al., Alcohol-based hand sanitizers: does gelling agent really matter? Gels 8 (2) (2022) 87.

[11] B.R.B.D. Costa, L.P.E. Haddad, V.L. Caleffo Piva Bigão, B.S.D. Martinis, Quantifying ethanol in ethanol-based hand sanitizers by headspace gas chromatography with flame ionization detector (HS-GC/FID), Journal of AOAC International 105 (1) (2022) 11−18.

[12] V. Andal, R. Lakshmipathy, D. Jose, Effect of sanitizer on obliteration of SARS−CoV2/COVID 19: a mini review, Materials Today: Proceedings 55 (2021) 264−266.

[13] S. Arfania, A.R. Sayadi, M.R. Khalesi, LCD digital signage 21.5 inch automatic induction hand sanitizer billboardsfor flotation machines, Journal of the Southern African Institute of Mining and Metallurgy 117 (1) (2017) 89−96.

[14] Y. Eddy, M.N. Mohammed, I.I. Daoodd, S.H.K. Bahrain, S. Al-Zubaidi, O.I. Al-Sanjary, et al., 2019 Novel coronavirus disease (Covid-19): smart contactless hand sanitizer-dispensing system using IoT based robotics technology, Revista Argentina de Clínica Psicológica 29 (5) (2020) 215.

[15] H. Sathwara, P. Vaghela, S. Joshi, AROGYAKAVACHAM-automatic hand sanitizer dispenser with temperature measurement, Proceedings of the International e-Conference on Intelligent Systems and Signal Processing, Springer, Singapore, 2022, pp. 785−796.

[16] V. Tiwari, S. Kalamdani, R. Raut, P.K. Rajani, Hygieia: smart health and sanitizing dispenser, ICT Analysis and Applications, Springer, Singapore, 2022, pp. 335−343.

[17] A. Gupta, R. Kumar, Novel design of automatic sanitizer dispenser machine based on ultrasonic sensor. IEEE, ISSN, (0932−4747) (2020) 228−233.

[18] Maker.io Staff. Make an automatic hand sanitizer dispenser using Arduino, <https://www.digikey.com/en/maker/blogs/2020/make-an-automatic-hand-sanitizer-dispenser-using-arduino>, 2020.

CHAPTER FOURTEEN

Design and modeling of a novel blood sampling (Phlebotomy) chair

Dilber Uzun Ozsahin[1,2,3], Basil Bartholomew Duwa[3,4], David Edward[6], Majd Issam Ali[4], John Bush Idoko[5] and Ilker Ozsahin[3,7]

[1]Department of Medical Diagnostic Imaging, College of Health Science, University of Sharjah, Sharjah, United Arab Emirates
[2]Research Institute for Medical and Health Sciences, University of Sharjah, Sharjah, United Arab Emirates
[3]Operational Research Center in Healthcare, Near East University, Nicosia/TRNC, Mersin 10, Turkey
[4]Department of Biomedical Engineering, Near East University, Nicosia/TRNC, Mersin 10, Turkey
[5]Applied Artificial Intelligence Research Center, Department of Computer Engineering, Near East University, Nicosia/TRNC, Mersin 10, Turkey
[6]Department of Biochemistry, Faculty of Basic Medical Sciences, University of Jos, Jos Nigeria
[7]Department of Radiology, Brain Health Imaging Institute, Weill Cornell Medicine, New York, NY, United States

Contents

14.1 Introduction

Blood sampling is a critical process in the medical field and it depends on how diseases are diagnosed through medical lab examinations. The process is conducted by drawing blood and placing the drawn blood in special tubes for further medical examinations. This is the first step in the process of diagnosing diseases in the medical laboratory department. It

Practical Design and Applications of Medical Devices.
DOI: https://doi.org/10.1016/B978-0-443-14133-1.00015-X

is a simple and quick process, and at the same time, it is considered one of the most important processes for diagnosing diseases. Specialists sometimes face some difficulties in drawing the blood sample because of the difficulty in accessing the veins from which the blood sample is drawn [1−4]. There are many cases and common mistakes that occur when the blood is drawn from the patient by the doctor while performing the blood drawing process, which negatively affects the results of the medical examinations. As such there are rejected samples that occur as a result of an error in the drawing process due to the unclear location of the vein from which blood is drawn. This error causes what is called decomposing samples (hemolysis samples or clotted samples) [2]. For example, when the doctor performs the procedure to draw blood from the patient's vein that is not clear or difficult to feel by the doctor, there is a possibility of red blood cells being broken due to the force of drawing the syringe in the process of drawing blood (the exit of blood from the patient is weak due to the wrong entry into the vein, which causes a state of breakage in the red blood cells) [5].

The slow process of drawing blood in the event of an inaccurate injection results in clotted blood samples. It greatly affects the reading of many medical laboratory examinations and gives incorrect results for the patient's condition and diagnosis, for example, in a clotted case, it gives a hemolysis result, resulting in blood platelet deficiency in the CBC examination [6−8]. Due to previous cases of wrong readings and diagnoses in laboratories, the intravenous detection device appears important to avoid the occurrence of such cases and errors, which in turn helps to unveil the vein from which the doctor wants to draw blood and give it for correct diagnosis. On the other hand, the lack of clarity of the location of the vein may lead to damage in the arteries and the capillaries surrounding it as a result of repeated punctures during the process of blood drawing [6].

The new blood drawing chair is made up of the following components: a chair that can be changed to a bed (to provide comfort to the patient), a sensor to detect the patient's temperature and blood pressure automatically, a vein detection device to perform laser vein detection, comfortable handles, and wheels to move the chair to transport the patient to the emergency room in urgent cases, such as a stroke or coma [1]. Generally, a blood test is performed with the person's arm extended and placed upside down on a flat surface or table. The examiner tightens an elastic belt around the upper arm to temporarily block blood flow in the hand and trap blood in the area from which the sample is taken. The person undergoing the examination is then asked to clench the fist and then slowly release the fist, relaxing the hand. This helps the person performing the examination find the best blood vessel, which is usually a vein in the elbow or forearm area. In some cases, a biopsy is taken from the back of the palm. When a suitable vein is found, the site is disinfected and sterilized with a rag infused with alcohol, and then a fine needle is inserted into the vein. This needle is attached to a test tube or a special syringe. This process may not be gentle and comfortable since it causes a slight pain when pricking. After this, the specialist performing the test draws the required amount of blood based on the number of test tubes.

When the blood drawing process is completed, the needle is quickly removed from the vein, and the puncture region is held on to prevent blood flow. Test tubes are taken to the laboratory for testing. In some cases, two or three attempts at pricking are needed, when vein penetration is not possible, or when the person performing the examination is unable to take the required amount of blood the first time. The process is not always simple and often takes between 5 and 10 minutes.

14.2 Materials and method

The force exerted on the blood vessels is expressed as blood pressure. The regularity of blood pressure is important to ensure the transport of oxygen and nutrients in the blood, in addition to hormones, white blood cells, and antibodies important to the body's immunity, to different parts of the body [7]. Blood pressure is measured at two different numbers: diastolic and systolic. Diastolic blood pressure is the pressure on the arteries during the diastole of the heart muscle, while systolic blood pressure is the pressure on the arteries during the contraction of the heart muscle. Blood pressure is expressed in millimeters of mercury. This causes some people to suffer from a disturbance in the rate of blood pressure, which causes it to decrease or rise above the normal level.

In most cases, a blood pressure disorder, or hypertension, may not be accompanied by any clear symptoms in the affected person. As such this disease is called a silent killer. But in some rare cases the patient may suffer from headaches and vomiting when suffering from high blood pressure [9−12]. Measuring blood pressure is the only way to know if the value of blood pressure has risen above the normal rate. Low blood pressure may be accompanied by a number of symptoms, such as tiredness and fatigue, a sense of dizziness and lightheadedness, blurred vision, loss of consciousness, and nausea. The severity of these symptoms varies from person to person.

14.2.1 Electronic blood pressure monitor

There are many different blood pressure measuring devices, one of which is the electronic blood pressure monitor, which is considered the best due to its ease of use and accuracy compared with other blood pressure measuring devices. It is advised to use high-quality devices to avoid getting false readings that may lead to inappropriate treatment plan or drugs prescribed. The instructions attached to the pressure device must be followed carefully to obtain correct results, as the instructions for using an electronic blood pressure monitor may differ from one device to another. The interpretation of blood pressure measurement results varies from person to person depending on the person's health status. It is also affected by factors such as weight, age, physical activity, and sex. If the blood pressure results are abnormal, the measurement must be repeated after 5 minutes to ensure the correct reading [11]. In the event that an abnormal result continues, appropriate measures must be taken, and the following is a statement to explain some of the results of blood pressure.

14.2.1.1 Low blood pressure
If the diastolic blood pressure ranges between 40 and 60 mmHg, and the systolic blood pressure ranges between 70 and 90 mmHg, it is considered low blood pressure.

14.2.1.2 Normal blood pressure
Blood pressure is considered normal if the diastolic blood pressure ranges between 60 and 80 mmHg, and the systolic blood pressure ranges between 90 and 120 mmHg.

14.2.1.3 Hypertension
Blood pressure is considered high if the diastolic blood pressure is less than 80 mmHg, and the systolic blood pressure ranges between 120 and 129 mmHg. However, this elevation is always not sufficient for diagnosing high blood pressure [10].

14.2.1.3.1 First stage of high blood pressure
The first stage of high blood pressure is diagnosed if the diastolic blood pressure ranges between 80 and 89 mmHg, or the systolic blood pressure ranges between 130 and 139 mmHg.

14.2.1.3.2 Second stage of hypertension
The second stage of hypertension is diagnosed if the diastolic blood pressure is equal to 90 mmHg or more, or the systolic blood pressure is equal to 140 mmHg or more.

14.2.1.3.3 Crisis in high blood pressure
This condition is considered an emergency and calls for seeing a doctor immediately, and is represented by the diastolic blood pressure reaching 120 mmHg or more, or the systolic blood pressure reaching 180 mmHg or more [12].

14.2.2 Vein detector device

Patients around the world suffer from painful moments while undergoing blood tests, especially when doctors or specialists search for the vein to take blood samples for examination and are unable to find it. They repeat the process until they get the vein, which causes pain and swelling of the skin. To avert such pain during blood sampling, we decided to design a seamless and painless procedure, which is the proposed integrated system. The proposed device (vein detector) has been successful in facilitating blood sampling and reducing pain in patients from whom blood is drawn. The device makes it easier for the doctor to find the location of the vein and not repeat the prick more than once. The proposed device has been tested on some donors to draw blood and has proven its efficiency in blood sampling.

14.2.2.1 Working principle of vein detector
By using a specific wavelength of infrared light to illuminate the skin, blood vessels absorb light due to the presence of hemoglobin, while other tissues scatter strongly and absorb weak light. Infrared imaging devices are used to capture light reflections or scattered near the infrared, passing information, which is captured through electrophoresis and image processing, after which the final location of the blood vessels is shown on the screen. It is distributed over the skin to monitor the blood vessels in real time. Invisible infrared light does not harm the human body.

14.2.2.2 Advantages and use cases of the proposed device

The proposed device is easy to operate and use. It has an independent battery design, with this, the device can work continuously for 1.5 hours after charging. It adopts AC and DC power supplies, which can be flexibly selected according to the environment, and has a low battery prompt. The device has a wide distance working design to avoid frequent height adjustments during use. The background color can be turned on by one button, and the user can set it according to their preference. The HD DLP projection screen is of high contrast. The high-speed digital FPGA processing platform produces high-definition real-time images.

14.2.3 Temperature and measurement

Temperature is a manifestation of thermal energy, and it is a physical property of matter that quantitatively expresses how hot or cold an object is. Temperature determines the direction of heat transfer. Temperature is measured by a thermometer, and the standard temperature scales are Celsius (degrees Celsius), Fahrenheit, and Kelvin. Although Kelvin is the standard used in the International System of Units to measure temperature, Celsius is the most common standard. The human body temperature ranges between 36.5°C and 37.5°C, and the normal body temperature should be around 37°C. According to the classification of the World Health Organization, 32°C is considered a severe decrease, while 38°C or more is a severe increase in body temperature. Remote measurement of temperature is done by an infrared device that gives a measurement of the temperature only, and it is called a point radiometer or an infrared thermometer [9]. It is effective in detecting temperature in people with a fever (a higher-than-normal body temperature) if used correctly.

14.2.3.1 Features of the proposed device

The proposed device does not emit any harmful radiation into the patient's body. It does not require the use of any chemicals. It gives immediate results that can be analyzed immediately. The data can be recorded and saved on the device's memory directly. The device is easy to carry and use and measures the ambient temperature from 0 to 100 degrees. The measurement distance for the body ranges from 5 to 15 cm.

14.2.3.2 Sources of error in infrared (contactless) body temperature measurement

There are many errors that can greatly affect the accuracy of body surface temperature readings. The most important potential sources of error associated with temperature measurement may come from the human body, the surrounding environment, or the thermometer itself.

14.2.3.2.1 Human body

There are several factors that can lead to significant changes in body surface temperature, including circulation problems, infections, and the use of certain medications. Sweat or surface moisture can also lead to a decrease in body surface temperature, and physical activity and the use of stimulants, including caffeine and nicotine, are all capable of increasing body surface temperature. In general, the normal temperature changes in the range of 0.6°C during the day, depending on the extent of human activity. It is not constant during the day and changes depending on the time at which the measurement was taken. In addition, the body temperature is sensitive to hormone levels [6].

14.2.3.2.2 The surrounding environment

The ambient air temperature can cause significant changes in the temperature of the human body. For example, measuring a person's temperature while in a hot car will give an unrealistic high-temperature reading. In addition, hot or cold air currents can cause large changes in temperature as well. Finally, central heat or cold sources can also raise or lower the body's surface temperature.

14.2.3.2.3 Temperature scales

Temperature gauges are not self-diagnostic. In addition, due to the lack of standardization among the manufacturers of these devices, significant differences in performance can result between different brands and models of equipment. Also, if there is any malfunction in the sensors, the device will read the wrong way. This device can measure the temperature remotely, which was of great help during the Covid-19 pandemic. Patients with high temperatures are tested for Covid-19. This device can be used in emergency situations and can be moved from one place to another.

14.2.4 Components of the blood drawing device

14.2.4.1 Movable chair

It is a movable wheelchair (similar to that a dentist uses), which can be converted into a bed in cases of emergency (in cases of coma). The parts of the chair are discussed in the following sections.

14.2.4.1.1 Chair base

The base of the chair is of an appropriate size to suit all sizes and ages.

14.2.4.1.2 Back of the chair

The back of the chair is flexible so that the chair can be converted into a bed, which is attached to a headrest for the comfort of the patient.

14.2.4.1.3 Footrest

The footrest is movable so that the chair can be converted into a bed in emergency situations.

14.2.4.1.4 Moving joints

The joints are capable of movement so that the doctor can convert the chair into a bed in an emergency, and the chair can be controlled using buttons fixed on the hand rests.

14.2.4.1.5 Strong iron structure

This chair consists of a strong iron structure that can withstand the most difficult conditions. The iron structure is hollow, so that it is lightweight and can be moved easily. The structure covers a layer of compressed sponge to ensure the greatest comfort for the patient. The sponge is covered with a layer of leather to facilitate its cleaning and sterilization and to prevent the transmission of diseases, as leather can sustain a lot of pressure compared with normal fabric.

14.2.4.2 Wheels

This machine consists of eight wheels:
- Four wheels connected to the base
- Two wheels attached to the back of the chair, which can be used and opened in emergency situations
- Two wheels connected to the footrest can be opened and closed when needed.

These are double wheels to ensure protection so that the chair does not tip in the event of a broken wheel. These wheels are connected by joints that can move 180 degrees [1].

14.2.4.3 Hand rest

The device consists of two armrests that are connected to the joints. They can be moved so that the patient can sit on the chair easily. It also has a function to install special buttons to move the chair and transform it into a bed [3]. The frame of the armrest is made of iron, and the structure is covered by a layer of pressurized sponge and then covered with a layer of leather.

14.2.4.4 Battery

The device contains a large-capacity battery to help us use the components of the device in the event of a power outage, or in the event that we want to transfer the device from one place to another. The battery is connected to a cable that can be connected to electricity to charge it when the power becomes low. The battery location is below the chair and is isolated by its material from any external influence to protect and preserve it.

14.2.4.5 Advanced display screen

It is a screen divided into several sections, which displays the patient's temperature and pressure. It also shows the malfunctions that may occur in the device so that they are easy to fix.

14.2.4.6 A vein detector

It is the primary device that detects the patient's veins, from which blood is drawn. The detector is carried by an iron arm, and it is connected by joints that can move in all directions to direct the rays to the patient's hand, from which the blood is to be drawn correctly. The vein detector can be connected to the screen to have an enlarged view of the patient's veins.

14.2.4.7 Cable with pressure sensor

It is a cable connected to a piece that is placed on the wrist of the patient's hand that measures the degree of pressure on the patient's body by sending signals through the cable connected to it to the display screen installed on the device. The result of the pressure measurement appears on a screen in front of the doctor.

14.2.4.8 Sensitive for remote measurement of temperature

A sensor is installed on one of the hand rests that measures the temperature of the body. This sensor is connected to a cable and sends signals to the screen, which displays the results of the examination of the temperature of the patient's body.

14.2.4.9 Cable to connect the device to electricity

This cable has two functions, which include connecting electricity to the device in case the device is installed in hospitals and clinics, and charging the battery so that we can use the device in the event of a power outage. The cable is wide to withstand the electrical pressure and is connected to a fuse. In the event that there is excessive current on the device, the fuse burns out and the device does not suffer any damage.

14.2.5 Working principle of blood drawing device

The patient sits on the chair, and then the doctor takes the appropriate position for the patient so that he moves the chair through the buttons fixed on the armrest and then installs the special piece to measure blood pressure. When the patient puts his hand on the armrest, the temperature sensor measures the patient's temperature automatically. Then the readings (pressure and temperature) appear on the screen installed on the proposed device.

The doctor directs the vein detection device to the appropriate part of the patient's body to perform blood sampling, so that the vein detector shows the patient's veins from which blood will be drawn. The doctor chooses the appropriate vein and injects the needle to draw the blood.

In emergency situations (comas or cases that require rapid surgical intervention), the chair is stretched to convert it into a bed (through the buttons fixed on the armrests) so that the patient can be transferred easily to the operating room.

14.3 Conclusion

This paper introduces a great idea in blood sampling to mitigate pain and difficulties faced by patients in the process. The idea also helps doctors experience a short turn time while performing blood sampling.

The proposed device is very helpful in emergency situations, such as cases of coma, and cases that require surgical intervention (because the chair can be transformed into a bed and can be moved from one place to another without the need to dispense with additives due to the presence of a rechargeable battery). The performance of the proposed integrated system is greater compared with when the integrated devices work separately.

References

[1] I. Lavery, P. Ingram, Blood sampling: best practice, Nursing Standard 19 (2005) 55–65.
[2] H. Galena, Complications occurring from diagnostic venepuncture, Journal of Family Practice 34 (5) (1992) 582–584.
[3] B. Newman, et al., The effect of whole-blood donor adverse events on blood donor return rates, Transfusion 46 (2006) 1374–1379.
[4] B. Cullen, et al., Potential for reported needlestick injury prevention among health-care personnel through safety device usage and improvement of guideline adherence: expert panel assessment, Journal of Hospital Infection 63 (2006) 445–451.
[5] L. Berkeris, et al., Trends in blood culture contamination. A college of american pathologist q-tracks study of 356 institutions, Archives of Pathology and Laboratory Medicine 123 (2005) 1222–1226.
[6] M. Little, et al., Percutaneous blood sampling practice in a large urban hospital, Clinical Medicine 7 (2007) 243–249.
[7] G. Mancia, R. Fagard, K. Narkiewicz, J. Redon, A. Zanchetti, M. Bohm, et al., 2013 ESH/ESC Guidelines for the management of arterial hypertension: the task force for the management of arterial hypertension of the European society of hypertension (ESH) and of the european society of cardiology (ESC), Journal of Hypertension 31 (2013) 1281–1357.
[8] C. Sierra, A. de la Sierra, Early detection and management of the high-risk patient with elevated blood pressure, Vascular Health and Risk Management 4 (2008) 289–296.
[9] J. Handler, The importance of accurate blood pressure measurement, The Permanente Journal 13 (2009) 51–54.
[10] H. Asada, A. Reisner, P. Shaltis, D. McCombie, Towards the development of wearable blood pressure sensors: a photo-plethysmograph approach using conducting polymer actuators, in: IEEE-EMBS, Proc. 27th annual international conference, Shanghai, China, pp. 4156–4159, 2005.
[11] K.W. Chan, K. Hung, Y.T. Zhang, Noninvasive and cuffless measurements of blood pressure for telemedicine, in: IEEE-EMBS, Proc. 23rd annual international conference, Istanbul, Turkey (4) 3592–3593, 2001.
[12] S. Colak, C. Isik, Blood pressure estimation using neural networks, in: Proc. IEEE international conference on computational intelligence for measurement systems and applications, pp. 21–25, 2004.

CHAPTER FIFTEEN

A speech recognition system using technologies of audio signal processing

Dilber Uzun Ozsahin[1,2,3]**, Declan Ikechukwu Emegano**[3,4]**, Abdulsamad Hassan**[4]**, Mohammad Aldakhil**[4]**, Ali Mohsen Banat**[4]**, Basil Bartholomew Duwa**[3,4] **and Ilker Ozsahin**[3,5]

[1]Department of Medical Diagnostic Imaging, College of Health Science, University of Sharjah, Sharjah, United Arab Emirates
[2]Research Institute for Medical and Health Sciences, University of Sharjah, Sharjah, United Arab Emirates
[3]Operational Research Center in Healthcare, Near East University, Nicosia/TRNC, Mersin 10, Turkey
[4]Department of Biomedical Engineering, Near East University, Nicosia/TRNC, Mersin 10, Turkey
[5]Department of Radiology, Brain Health Imaging Institute, Weill Cornell Medicine, New York, NY, United States

Contents

15.1 Introduction

Rehabilitation services, according to the World Health Organization (WHO, 2021) [1] are necessary for about 5% of a country's population, or 430 million individuals, to deal with their "severe disability" hearing damage (432 million adults and 34 million children). It is anticipated that in the year 2050, nearly 700 million individuals, or one out of every 10 people, would have hearing loss that is severe enough to be disabling [1]. WHO further

Practical Design and Applications of Medical Devices.
DOI: https://doi.org/10.1016/B978-0-443-14133-1.00001-X

reiterates that hearing loss that is considered disabling is defined as having a severity that is more than 35 decibels in the ear that is better able to hear. Approximately 80% of people who have severe hearing problems suffer enough to impair their ability to function effectively, especially in low or intermediate nations. The incidence of hearing problems grows with aging; among people 60 and older, more than 25% are impacted by hearing loss severe enough to impair their ability to perform daily activities. WHO estimates that the global yearly cost of untreated hearing loss is 980 billion US dollars. This comprises costs incurred by the health sector (except the costs associated with hearing aids), costs incurred by education services, costs associated with loss of productivity, and societal expenditures. It is estimated that low- to middle countries are accountable for 57% of these expenditures [1]. The speech recognition system identifies spoken words in a speech and converts them into written text. They are sometimes called automatic speech recognition and the device is usually used per person [2]. Hearing losses are caused by several factors, ranging from genetics to infection to accidents, viral outbreaks, and noise blasts. [3] Hearing disorders are one of the main problems that people suffer from in today's world. These disorders range from partial, moderate, severe, or complete lack of hearing (deafness) [4]. This disorder can affect one or both ears and results in total hearing loss of conversations or sounds. Persons with mild-to-severe hearing loss are hard of hearing [5]. They often use sign language as a medium of communication. Hearing disorders are treated in many ways medically, such as through surgery, or using assistive devices to improve hearing. The appropriate treatment depends on the patient's condition [6]. So, by designing a voice recognition system that converts human speech into written text, speech recognition systems use multiple mathematical algorithms to process the audio signal and extract the characteristics that are useful in recognizing the desired word [7]. Assistive gadgets are provided by the technology that is being developed to combat this issue. Individuals who are hard of hearing can benefit more from their training when it makes use of modern technologies [8]. Among these technologies are speech recognition devices. Speech recognition extracts sound waves in the form of acoustic signals from speech. These words are determined according to the machine patterns [1−3]. The ability of a programmed machine to recognize and identify words spoken by patients and convert these words into text that can be read requires software that does the job perfectly. Recognition software is limited to a small vocabulary, and therefore the identification of words and phrases spoken clearly will be retarded [9]. This speech-to-test system is used mostly by people with hearing disorders or impairments.

15.1.1 Categories of a speech device

The categories of speech devices that are in vogue are front-end speech recognition technology (SRT) systems, back-end-SRT systems, systems designed for multiple users and small-vocabulary users, and systems designed for limited users and large-vocabulary users [10]. Speech recognition systems are also classified according to the nature of speech [11]. The speech recognition process used in the search goes through many stages until the word appears in written text on the screen. A reference library is initially built that includes several audio recordings of several talks with the characteristics extracted from these recordings. However, alternative approaches and algorithms, such as principal component analysis, the Address Resolution Protocol, and the Mel frequency cepstral coefficient (MFCC) [12] can be used to analyze audio signals. The first among the studies looked at designing a hardware control system through speech recognition using MATrix LABoratory (MATLAB®) [13].

15.2 Related studies

M. Shankar-Hari et al, [14] in their work studied the performance of the MFCC and DTW (dynamic time warping) for bird sound classification. The researcher uses lovebirds (Agapornis), which are popularly used as pets [14]. The sound of a lovebird whistling could be learned. It could also be used to recognize a phase of the experience of learning speech recognition. Speaker recognition takes note of Agapornis's voice frequency and then relates it to the frequency of sounds in the data sets [14,15]. The findings indicate that sound validation had an average accuracy of 80%. According to the results of previous analyses, designs, as well as discussions, MFCC plus dynamic warping time can be used to make sound categorization applications that can tell how good lovebirds are [15]. The study also uses isolated words with constant filters in trigonometric MFCC as a way to extract the characteristics of sound recordings of speech and deformation time [16]. In addition to this, the works of other researchers utilize similar algorithms. A database of five people was built [17], and each of them logs ten different words, which were constantly extracted by filtering trigonometric recordings and then using an algorithm known as DTW. This method is based on a measure of similarity between the two chains of change in time and speed [18]. From the previous articles, it was an easy system to construct due to the application's simplicity and the reference library's modest number of terms. More so, in the field of bandwidth, the system is also characterized by low computing power consumption for implementing the identification process [19]. However, due to the simplicity of the methods and algorithms employed in the identification process, this system can only grasp and know words and terminology in English, and any auditory interference when performing a command entry to be identified may lead to misunderstanding [20], even if the voice command is correct and matches the commands contained in the reference library. There is also a noise signal that prevents identification [21]. However, in one of the previous studies by M. Shankar-Hari, et al, a hardware control system was designed through speech recognition using MATLAB. The researchers used the Euclidean distance measurement method for speech recognition by comparing the samples of the input signals with the reference samples and returning the models that give the highest percentage of similarity, so the voice command is determined, and then the system executes the required voice command [22]. Another previous study by S. Dev Dhingra et al identified isolated words using MFCC trigonometric filter constants and DTW busy time distortion. These methods measure the similarities between two series characterized by a change in velocity and time [23]. Also, many studies in the past have used neural networks in speech recognition systems. Neural networks help the user design the method to fit their needs [24]. All the studies are very scientific indeed, notwithstanding this novel and unique study that entails the use of a software electronic system in designing a speech-defined system used in hardware control operations.

15.3 Methodology

15.3.1 Sound transmission and reception

Sound waves travel through a variety of mediums, such as solids, liquids, and gases. It is only sensed by the ear in a vacuum. Sound signals are also generated by an organism. The organism with the source organ of sound uses it to communicate with another organism of its type or gender, consciously or unconsciously, and is called the source organ [25]. These vibrations cause the sensation of hearing at a speed of approximately 340 m/s [26]. The viscosity and magnetic field influence through which the sound travels determine the sound wave as it moves away from the source of the audio [27]. The characteristics of sound are wavelength, i.e., the distance between two peak levels measured in centimeters, meters, micrometers, or even nanometers (nm) [28]. The frequency (f) is the number of times a specific wave repeats in a certain amount of time. The regularity, or frequency, f, is equal to one-half of the time, t. This is shown by the equation $f = 1/t$. The frequency is measured in hertz. Sound waves have an amplitude (A), which is the altitude of the wave. The intensity (f) and the distance of a wave can be used to figure out its speed. Sound volume, on the other hand, shows how loud the sound is at a certain point, which is equal to the amount of energy per unit in the right-angle area of the wavefront whose center is that point. Sound intensity is measured in watts per m^2 [29]. The classification of sound wave is dependent on these characteristics.

15.3.1.1 Audible waves

This has a frequency of between 20 Hz and 20 kHz, which is the frequency of human hearing. However, in the elderly, it is reduced to 12 kHz. The maximum a human ear can hear is from a range of 5000 Hz to 8000 Hz, inclusive of letter vibrations. People's vocal cords and musical instruments can both make sound waves [30]. Human ears do not detect waves of greater than 20 kHz, though this is still under scientific research [31]

15.3.1.2 Infrasound waves

They are infrasonic waves or vibrational sound waves at frequencies lower than 20 Hz that cannot be heard, which originate from the seismic and slip movements of the earth's crust. This is the reason it is vital to monitor the activities of volcanoes and earthquakes [32].

15.3.2 Different types of recognition systems

The different recognition systems are front-end SRT, back-end SRT, and continuous-speaking and discrete-speaking recognition systems. In the front SRT, the users dictate speech to a computer using a microphone. Then a process is performed to convert words into written text using processing applications in real time. To improve the accuracy of the recognition process, the user must quickly correct the errors that the program may make. So that the program learns and preserves the user's appearance and modern style with better efficiency. However, the back-end SRT recognizes speech and converts it into text. The speech takes place after the user dictates the speech on the computer; that is, systems that work in real-time. In these systems, dictation is first done, and then the words are converted into a digital image. The digital image is then converted into a written text [33]. This is limited by very different word patterns spoken by people, ways of pronouncing words, and the shape of their voice signals that affect the digital image [33]. Modern systems are designed to have extraordinary linguistic knowledge and relationships in eradicating this shortfall. Modernized systems have an accuracy of 58% accuracy, though this can be increased depending on the user's experience. They are named system-based users [34]. Another speech algorithm system is the continuous-speaking and discrete-speaking recognition system [35].

15.3.3 Speech recognition algorithm

A digital system detects the voice and prints it on the patient's computer screen so that the patient knows exactly what is going on. Only voice recognition is used in these processes. A human voice, which is an audio signal, is turned into textual content by a programming language that records it and figures out what it means. Speech recognition systems utilize computational models to pull information about the person speaking out of sound waves and start comparing it to the information stored in a source file for the system or programming language. So, the voice is picked up. To figure out what a word or sentence is, a comparative analysis with the loaded library is needed. In this study, a software electronic system is designed to build a speech-defined system used in hardware control operations. The setup consists of an electronic board that functions as an audio signal using a microphone. Then the audio signal filtration is set between 300 Hz and 4000 Hz to get the voice signal frequencies. This is preconfigured using a computer system via RS-232 and processed and recognized through the MATLAB algorithm [36]. The computer program with the aid of MATLAB implements an algorithm known as a library of recognizable words, and then the unknown words are applied. This device was tested for similarity using distance measurements on the scale of Euclidean [37] and was compared. Once the process is identified, the user can issue arbitrary orders through the system's voice portal [38].

Cepstrum analytical techniques are applied to remove noise barriers [39]. After that, the user recognizes the spoken words using a speech recognition algorithm that compares the characteristics extracted from the input audio signal with the attributes in the standard library and selects the word that achieves the highest similarity criterion [40]. Finally, the system is trained to obtain the best possible results. From Fig. 15.1 of the outline, the speech is made at the voice signal. This voice is processed and extracted, which the algorithm recognizes. This algorithm detects if the spoken words match the library of words, which are then processed and displayed in the output as written words automatically.

This device has similar characteristics to that of the electrocardiogram signals in terms of amplitude, basic forms, and frequency [41]. The basic explanation for these differences in the phonetic signals is that they are the result of essential factors playing basic roles. More so, the transformation of speech into processed data is never a procedure but rather requires a better algorithm that has expanded properties capable of distinguishing an individual speech from other speeches. To convert these speeches into processed data, we need to convert the audio signals to digital signals, carry out primary processing operations to improve the signals, and finally implement signal-specific algorithms to extract valuable properties from the audio signals. This is simplified in Fig. 15.2.

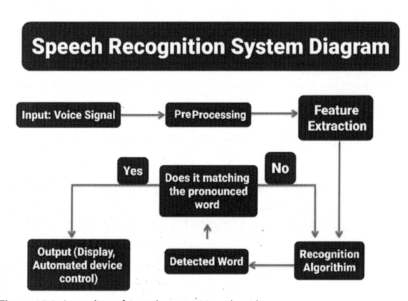

Figure 15.1 An outline of speech recognition algorithm.

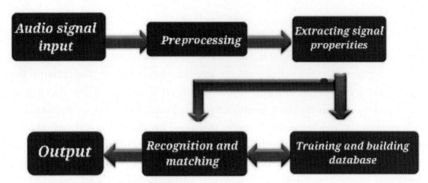

Figure 15.2 Intended circuit diagram.

Audio signal input expressing human speech must be entered. The input signal is used to form the system's library, where the characteristics are extracted and saved. The input signal is then used in the identification process, comparing its properties to those stored in the reference library. To convert the audio signal, it is necessary to convert the signal from its analog form to a digital format within the process. This is done by choosing a sampling frequency suitable for the audio signal, i.e., the number of frequencies per second in a continuous-discrete signal. The sampling frequency is harmonized with the clipping frequency for all audio samples entered [42,43] whether for the reference library or identification. The frequency is usually 8000 Hz so that it can accommodate the human audio signal's range (3600 Hz−300 Hz) and is in the Nyquist limit [44,45].

This signal moves in a wavelike manner, though it cannot be seen with the naked eye. The primary audio signal process aims to improve the signal quality to facilitate the performance of subsequent processing steps as well as the accuracy of the extraction for the best possible form. During this initial signal, three basic things happen: (1) The audio signal uses a low-filter pass, (2) Divides the audio signal and reads it in time frames of equal length, and (3) Finally deletes the silent times in the audio signal. In Fig. 15.3 the yellow blocks are the low pass filters. The setup is tested by saying the word "hello" using the wave recording functions for 3 seconds at a cut-off frequency of 8000 [Hz], as can be seen in Fig. 15.4

The audio signal is split and then read into equal-length time frames to facilitate extraction. About 20 milliseconds was chosen since it is the shortest value at which a phoneme can exist, and choosing time frames shorter than this will be meaningless and lead to additional computational problems. In other words, the audio signal is steady at 21 milliseconds

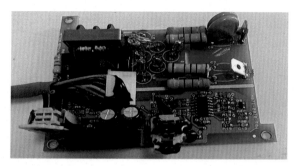

Figure 15.3 Audio signal input.

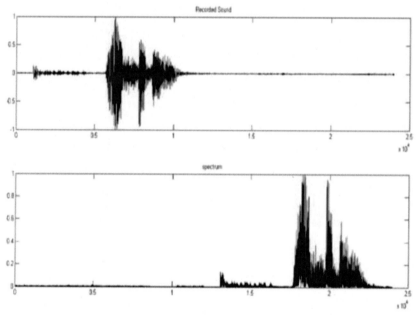

Figure 15.4 Signal generated by recording the word "hello" using the wave recording function.

because it is regarded as constant at these values. The final process is the preprocess function that deletes quiet moments in the audio signal.

15.3.4 Recognition results

The audio output drives the digital signals into its output. It has Arduino and pulse-duration modulation, as can be seen in Fig. 15.5, for the recognition process. The system succeeded in recognizing all the words entered into it compared with the properties stored in the standard library. The

Figure 15.5 Audio output.

recognition library could be developed using five audio recordings of two separate people under specific conditions, such as a calm recording environment and no extraneous noise. The prior findings were only for a few words, indicating that the recognition procedure was 100% accurate. A further study could expand the findings by increasing the number of words and recordings for each word, recording the findings, and calculating the recognition percentage. The system returns the word with the most similar qualities to the given word, therefore if a new word is entered in the properties library, the system returns the closest term (hidden Markov model). The system will always compare the characteristics of the entered audio recording with the characteristics saved in the standard toolbox, and return the highest similarity between them (Fig. 15.6).

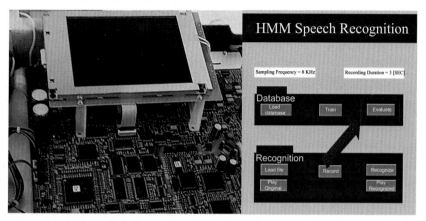

Figure 15.6 Digital display of the speech.

15.4 Conclusion

Speech is the most significant signal that we send and receive. Speech is the medium via which all of life's essential information can be conveyed. It is also the preferred form of communication whenever we wish to communicate with another person, with a few notable exceptions, including certain disabilities. Therefore any technological advancement that we can make to facilitate communication patterns or that makes use of the fact that we can speak is potentially valuable. In a similar vein, the development of enhanced web applications that achieve results in voice input enables more natural contact with electronic gadgets. The MFCC is representative of the automated classification of speech, and it is the feature that is utilized most frequently in speaking and sound recognition, as well as in classification techniques. The system can be used as an assistant for patients with hearing disorders, as it can identify isolated words and show them on the screen for patients with hearing disorders, including patients with complete deafness, providing them with an excellent way to communicate with the outside world, not wholly, but for the primary and essential words that benefit them in knowing what is going on around them. The study is emphatically for learning purposes. Therefore the design cannot replace the conventional audio devices used in the health sector.

References

[1] World Health Organization, Deafness and hearing loss. <https://www.who.int/news-room/fact-sheets/detail/deafness-and-hearing-loss>, 2023 (accessed 27.08.22).

[2] R. Pahwa, H. Tanwar, S. Sharma, Speech recognition system: a review, International Journal of Future Generation Communication and Networking, 13 (2020) 2547-2559. <https://www.researchgate.net/publication/343934770_Speech_Recognition_System_A_review> (accessed 15.08.22).

[3] V.N.(Mrs). Nwadinobi, Chapter eight hearing impairment. <https://www.researchgate.net/publication/336025368_CHAPTER_EIGHT_HEARING_IMPAIRMENT>, 2019 (accessed 27.08.22).

[4] C.S. Brown, S.D. Emmett, S.K. Robler, D.L. Tucci, Global hearing loss prevention, Otolaryngologic Clinics of North America 51 (3) (2018) 575–592. Available from: https://doi.org/10.1016/J.OTC.2018.01.006.

[5] T.C. Lin, M. Yen, Y.C. Liao, Hearing loss is a risk factor of disability in older adults: a systematic review, Archives of Gerontology and Geriatrics 85 (2019). Available from: https://doi.org/10.1016/J.ARCHGER.2019.103907. Nov.

[6] L. Lu, et al., Wearable health devices in health care: narrative systematic review, JMIR Mhealth Uhealth 2020 8 (11) (2020) e18907. Available from: https://doi.org/10.2196/18907, https://mhealth.jmir.org/2020/11/e18907. 8, (11), p. e18907.

[7] Z. Song, English speech recognition based on deep learning with multiple features, Computing 102 (3) (2020) 663–682. Available from: https://doi.org/10.1007/S00607-019-00753-0.

[8] B. Baglama, M. Haksiz, H. Uzunboylu, Technologies used in the education of hearing impaired individuals, International Journal of Emerging Technologies in Learning 13 (9) (2018) 53–63. Available from: https://doi.org/10.3991/IJET.V13I09.8303.

[9] L. Ben, K. Karolina, What is speech recognition?, Speech Recognition. <https://www.techtarget.com/searchcustomerexperience/definition/speech-recognition>, 2022 (accessed Jun. 24, 2022).

[10] S. Basma, B. Lord, L.M. Jacks, M. Rizk, A.M. Scaranelo, Error rates in breast imaging reports: comparison of automatic speech recognition and dictation transcription, American Journal of Roentgenology 197 (4) (2011) 923–927. Available from: https://doi.org/10.2214/AJR.11.6691.

[11] US5758023A, Multi-language speech recognition system, Google Patents. <https://patents.google.com/patent/US5758023A/en> (accessed 25.01.22).

[12] L.R. Rabiner, B.H. Juang, Fundamentals of speech recognition, PTR Prentice Hall, [WorldCat.org]. <https://www.worldcat.org/title/fundamentals-of-speech-recognition/oclc/26674087>, 1993 (accessed 24.01.22).

[13] B.D. Parameshachari, A study on smart home control system through speech, International Journal of Computer Applications. 2013 975–8887. <https://www.researchgate.net/publication/327530706_A_Study_on_Smart_Home_Control_System_through_Speech>, (accessed 02.07.22).

[14] M. Shankar-Hari, D.A. Harrison, P. Ferrando-Vivas, G.D. Rubenfeld, K. Rowan, Risk factors at index hospitalization associated with longer-term mortality in adult sepsis survivors, JAMA Network Open 2 (5) (2019). Available from: https://doi.org/10.1001/JAMANETWORKOPEN.2019.4900.

[15] H. Lu, L. Wang, N. Fang, N. Wang, Signal recognition method based on Mel frequency cepstral coefficients and fast dynamic time warping for optical fiber perimeter defense systems, Applied Optics 61 (7) (2022) 1758–1765. Available from: https://doi.org/10.1364/AO.448754.

[16] B. Milner, X. Shao, Prediction of fundamental frequency and voicing from mel-frequency cepstral coefficients for unconstrained speech reconstruction, IEEE Trans Audio Speech Lang Process 15 (1) (2007) 24–33. Available from: https://doi.org/10.1109/TASL.2006.876880.

[17] Z. Jeremy, Dynamic time warping: explanation and code implementation, Towards Data Science. <https://towardsdatascience.com/dynamic-time-warping-3933f25fcdd>, (accessed 02.07.22).

[18] H. Prapcoyo, B.P. Adhita Putra, R.I. Perwira, Implementation of mel frequency cepstral coefficient and dynamic time warping for bird sound classification, Conference SENATIK STT Adisutjipto Yogyakarta 5 (2019). Available from: https://doi.org/10.28989/senatik.v5i0.326. Nov..

[19] MIT Technology Licensing Office, Low power circuits for automatic speech recognition. <https://tlo.mit.edu/technologies/low-power-circuits-automatic-speech-recognition>, 2022 (accessed Jun. 25, 2022).

[20] R.J. Tanna, J.W. Lin, O. de Jesus, Sensorineural hearing loss, *NCBI Bookshelf*, pp. 1–13 [Online]. <https://www.ncbi.nlm.nih.gov/books/NBK565860/>, 2022 (accessed 13.07.22).

[21] W.J. Doedens, L. Meteyard, What is functional communication? a theoretical framework for real-world communication applied to aphasia rehabilitation, Neuropsychology Review 1 (2022) 1–37. Available from: https://doi.org/10.1007/S11065-021-09531-2/TABLES/1. Jan..

[22] H. Parmar, B. Sharma, Control system with speech recognition using MFCC and euclidian distance algorithm, International Journal of Engineering Research & Technology 2 (1) (2013). Available from: https://doi.org/10.17577/IJERTV2IS1384.

[23] S. Dev Dhingra, G. Nijhawan, P. Pandit, Isolated speech recognition using MFCC and DTW (2007). Available from: <http://www.ijareeie.com>.

[24] Xenonstack, Artificial neural networks applications and algorithms. <https://www.xenonstack.com/blog/artificial-neural-network-applications>, 2019 (accessed 25.06.22).

[25] A. Bell, B. Davies, H. Ammari, Bernhard riemann, the ear, and an atom of consciousness, Foundations of Science (2021) 1–19. Available from: https://doi.org/10.1007/S10699-021-09813-1/FIGURES/1.

[26] Study.com, The speed of sound in air is approximately 340 m/s. The speed of sound in steel is approximately 5900 m/s. If your friend strikes one end of a steel pipe with a hammer while you listen at the other end. <https://study.com/academy/answer/the-speed-of-sound-in-air-is-approximately-340-m-s-the-speed-of-sound-in-steel-is-approximately-5900-m-s-if-your-friend-strikes-one-end-of-a-steel-pipe-with-a-hammer-while-you-listen-at-the-other-en.html>, 2022 (accessed 02.07.22).

[27] Physics Tutorial: The Speed of Sound, Sound waves and music — Lesson 2 — sound properties and their perception. <https://www.physicsclassroom.com/class/sound/Lesson-2/The-Speed-of-Sound> (accessed 02.07.22).

[28] PASCO, Sound waves, PASCO Scientific. <https://www.pasco.com/products/guides/sound-waves> 2022 (accessed 13.07.22).

[29] M.A. Abdelaziz, D.G. Grier, Acoustokinetics: crafting force landscapes from sound waves, Physical Review Research 2 (1) (2020) 013172. Available from: https://doi.org/10.1103/PHYSREVRESEARCH.2.013172/FIGURES/5/MEDIUM.

[30] Byjus.com. Sound — audible and inaudible sounds. Frequency range and vibration. <https://byjus.com/physics/inaudible-audible-sound/> 2022 (accessed 25.06.22).

[31] M.D. Fletcher, S.L. Jones, P.R. White, C.N. Dolder, T.G. Leighton, B. Lineton, Effects of very high-frequency sound and ultrasound on humans. part i: adverse symptoms after exposure to audible very-high frequency sound, The Journal of the Acoustical Society of America 144 (4) (2018) 2511. Available from: https://doi.org/10.1121/1.5063819.

[32] G. Leventhal, What is infrasound? Progress in Biophysics and Molecular Biology 93 (1–3) (2007) 130–137. Available from: https://doi.org/10.1016/J.PBIOMOLBIO.2006.07.006.

[33] J. Bing-Hwang, F. Sadaoki, Automatic recognition and understanding of spoken language—a first step toward natural human-machine communication, Proceedings of the IEEE. 88. 1142 - 1165. 10.1109/5.880077. <https://www.researchgate.net/publication/2985738_Automatic_Recognition_and_Understanding_of_Spoken_Language-A_First_Step_Toward_Natural_Human-Machine_Communication>, 2000 (accessed 24.06.22).

[34] M. Sagar, K.S. Jasmine, Dialog management system based on user persona, Global Transitions Proceedings 3 (1) (2022) 235−242. Available from: https://doi.org/10.1016/J.GLTP.2022.03.029.

[35] E.L. Higgins, M.H. Raskind, Speaking to read: the effects of continuous vs. discrete speech recognition systems on the reading and spelling of children with learning disabilities, Journal of Special Education Technology 15 (1) (1999) 19−30. Available from: https://doi.org/10.1177/016264340001500102.

[36] L. Dhruv, Audio signal filtering − Rhea. <https://www.projectrhea.org/rhea/index.php/Audio_Signal_Filtering>, 2022 (accessed 12.07.22).

[37] P.A. Abhang, B.W. Gawali, S.C. Mehrotra, Technical aspects of brain rhythms and speech parameters, Introduction to EEG- and Speech-Based Emotion Recognition (2016) 51−79. Available from: https://doi.org/10.1016/B978-0-12-804490-2.00003-8.

[38] H. Chhatbar, T. Janak, C. Rahul, B. Darshan, Secure speech controlled robot using Matlab and Arduino, 02 (2015) 325−335. <https://www.researchgate.net/publication/344397188_Secure_Speech_Controlled_Robot_using_Matlab_and_Arduino> (accessed 25.06.22).

[39] MathWorks.com, Cepstrum Analysis − MATLAB & Simulink. <https://www.mathworks.com/help/signal/ug/cepstrum-analysis.html>, 2022 (accessed 25.06.22).

[40] A. Mukhamadiyev, I. Khujayarov, O. Djuraev, J. Cho, Automatic speech recognition method based on deep learning approaches for Uzbek language, Sensors 22 (10) (2022). Available from: https://doi.org/10.3390/s22103683.

[41] M. Murugappan, Prof, Frequency band analysis of electrocardiogram (ECG) signals for human emotional state classification using discrete wavelet transform (DWT), Journal of Physical Therapy Science 25 (2013) 753−759. <https://www.researchgate.net/publication/256493108_Frequency_Band_Analysis_of_Electrocardiogram_ECG_Signals_for_Human_Emotional_State_Classification_Using_Discrete_Wavelet_Transform_DWT> (accessed 15.08.22).

[42] F. Esqueda, S. Bilbao, V. Välimäki, Antialiased soft clipping using a polynomial approximation of the integrated bandlimited ramp function. <https://www.researchgate.net/publication/308693747_Antialiased_soft_clipping_using_a_polynomial_approximation_of_the_integrated_bandlimited_ramp_function>, 2016 (accessed 13.07.22).

[43] WHO, Addressing the rising prevalence of hearing loss, no. 02 (2018). Available: <https://apps.who.int/iris/bitstream/handle/10665/260336/9789241550260-eng.pdf?sequence = 1&ua = 1%0A http://www.hear-it.org/multimedia/Hear_It_Report_October_2006.pdf%0Afile:///C:/Users/E6530/Downloads/9789240685215_eng.pdf%0Ahttps://doi.org/10.1016/j.ijporl>.

[44] B. Vosooghzadeh, Issues in 5G Wireless Network: Network Slicing, Softwarization, Faster-Than-Nyquist Signaling & Signal Source Separation. 10.13140/RG.2.2.11148.16004. <https://www.researchgate.net/publication/330038664_Issues_in_5G_Wireless_Network_Network_Slicing_Softwarization_Faster-Than-Nyquist_Signaling_Signal_Source_Separation>, 2018 (accessed 13.07.22).

[45] W. Gavin, What is the Nyquist theorem?. Nyquist theorem. <https://www.techtarget.com/whatis/definition/Nyquist-Theorem>, 2022 (accessed 13.07.22).

CHAPTER SIXTEEN

Blood circuit in hemodialysis

Dilber Uzun Ozsahin[1,2,3], Declan Ikechukwu Emegano[3,4],
Bahaaeddin A.T. Bader[4], Basil Bartholomew Duwa[3,4] and
Ilker Ozsahin[3,5]

[1]Department of Medical Diagnostic Imaging, College of Health Science, University of Sharjah, Sharjah, United Arab Emirates
[2]Research Institute for Medical and Health Sciences, University of Sharjah, Sharjah, United Arab Emirates
[3]Operational Research Center in Healthcare, Near East University, Nicosia/TRNC, Mersin 10, Turkey
[4]Department of Biomedical Engineering, Near East University, Nicosia/TRNC, Mersin 10, Turkey
[5]Department of Radiology, Brain Health Imaging Institute, Weill Cornell Medicine, New York, NY, United States

Contents

16.1 Introduction

16.1.1 Blood circuits and hemodialysis

The kidney plays a vital role in the filtration, excretion, and metabolism of analytes. The kidney plays a role in hemostasis, regulation of water, and waste removal during urine production. The kidney is made up of millions of nephrons that serve as filtering units. The functions of the kidney cannot be underrated. The major function of the kidney is to

217

eliminate salts as well as harmful substances from the body, such as urea, creatinine, and other waste products, resulting from metabolic processes. These harmful substances leak into the blood and cause harm to the individual, resulting in kidney failure [1]. An artificial kidney device is used when a person has renal insufficiency, i.e., the natural kidneys do not function appropriately [2]. This device is not a replacement for natural kidneys, but it performs 60%−70% as optimally as normal kidneys [3]. Anatomically, kidneys are reddish-brown bean-shaped organs, with the size of a clenched fist, located at the posterior end of the abdomen [4] with the left kidney slightly higher in position than the right kidney. The length of each kidney is about 10 cm, its width is 5 cm, and its thickness is about 1.5 [5]. The kidney has a convex outer surface and a concave inner surface known as the umbilicus. The renal artery and vein enter the kidney and help blood to flow back into the heart [4,6,7]. The kidney is made up of the renal cortex, medulla, and pelvis. The renal cortex contains spherical granules known as Malpighi globules. There are more than 2 million of these globules in a human kidney. This also has an epithelial lining, i.e., urothelium, which starts as a monocellular layer and later expands up to six cell layers, the moment it reaches the ureter [8−10]. The kidney also has a minute tube that begins in the cortex area with a swollen double-walled part called Bowman's capsule, surrounded by a network of abundant blood capillaries called the glomerulus. There are about 1.2 million of these units in the human kidney, and each one has its own functional unit called the nephron. The Malpighian ball is the twisted tube that connects the cytoplasm to the kidney pelvis [11,12]. It is narrow and twisted because it is located near the Malpighian ball and in the area of the cortex. The collecting duct is a fine and straight tube in which the distal convoluted tubules flow. This tube unites with other grouped tubes to form larger and larger tubes, and these tubes finally flow at the top of the Malpighian pyramid [13].

16.1.2 Functions of the kidney

The major function of the kidney is filtration of metabolic waste from the bloodstream. The kidney also regulates electrolytes and excretes waste products, as well as contributes to blood cell production [14]. The acid–base balance of the body is also controlled by the kidney. In the human body, acids, and bases have a delicate balance, which is often reflected in the pH values. The normal value of pH in the blood is 7.35−7.45. For this to be maintained in equilibrium, the kidney plays a vital role in controlling the balance of water [13,15]. This is usually seen in the volume of voided urine during hydration and when the system is dehydrated. In addition, the electrolyte balance is maintained by the kidney [14]. The kidney's vital function is the removal of toxins and waste from the body [13]. The kidney filters water as well as soluble waste and toxins, which are flushed out from the urine. It converts angiotensinogen, which the liver produces as angiotensin 1. In the lungs, it is later converted to angiotensin II, which constricts the vessels, increasing blood flow [16]. More so, the kidney converts calcifediol to calciferol, which is the active form of vitamin D. The kidney plays a vital role in circulation, regulating calcium as well as phosphate balance in the body, which is needed for bone development [17].

16.1.3 Global mortality rates

In a study by V. Jha et al., kidney disease is reported to be one of the major causes of death globally. Hence, more than 2 million people are treated via dialysis, and about 10% of them need this form of treatment for survival [18]. Similarly, the death rates recorded for kidney-related diseases have drastically increased, especially in chronic kidney diseases (CKDs) compared with nonCKDs. In 2018, the mortality rate rose to 96 out of 1000 patients with CKDs compared with 41 persons out of 1000 patients with nonCKDs [19]. This mortality rate is dependent on the treatment of kidney failure. The mortality rate of patients on dialysis ranged from 15% to 20% within 5 years of survival, was attributed to 50%. About 80% of patients survive when their kidneys are transplanted [20]. Fig. 16.1 shows the impact of kidney diseases in some countries.

In Fig. 16.1, Micronesia has the highest mortality rate, followed by El Salvador. Egypt has the least number of deaths as a result of kidney disorders.

CKD is significantly more widespread globally. This type of kidney disorder affects about 10% to 15% of the adult population living in Western nations. The treatment of this disorder is very expensive. Sometimes it may require total kidney transplantation. Simultaneously, there is a rise in risk factors such as diabetes, hypertension, and even obesity [21]. In the 21st century, CKD has become one of the most prevalent causes of mortality, in part because of the increase in risk factors. In 2017, an average total of 843,6 million people around the world were diagnosed with CKD [22]. Although there is a decline in the death rate among end-stage renal patients, the burden of kidney disease globally, according to

kidney mortality rate in different countries

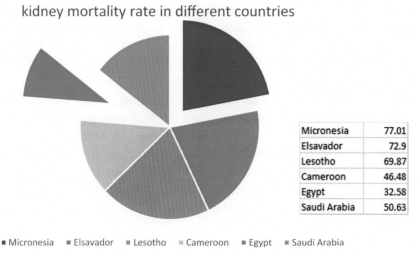

Micronesia	77.01
Elsavador	72.9
Lesotho	69.87
Cameroon	46.48
Egypt	32.58
Saudi Arabia	50.63

■ Micronesia   ■ Elsavador   ■ Lesotho   ■ Cameroon   ■ Egypt   ■ Saudi Arabia

Figure 16.1 Kidney mortality rate in different countries.

research, has increased and poses a significant concern for the general public. Therefore it is very important to make a diagnosis, monitor patients with this disorder, as well as treat those with dysfunctions. The global therapeutic techniques cut across transplanting and dialysis [23].

Statistics show that kidney diseases rank 12th among all diseases causing death in humans [24] and 17th among those that cause disability. According to the World Health Organization (WHO) records, it was responsible for 35 million deaths in 2005 [25]. When detected and treated early, it can stop getting worse and save money at the same time. In the United States, it is very costly to treat CKDs. The cost of treatment exceeds $48 billion per year [26]. In 2010, CKDs were the 27th leading cause of death around the world. Dialysis alone accounts for approximately 2 million people receiving treatment or kidney transplants worldwide. The major focus will be on the diseases that lead to hemodialysis [27,28]. In CKD, the kidney fails to keep the body healthy through the filtration of toxic substances from the blood [29]. As it progresses, waste is built up, causing nausea, anemia, nerve damage, and poor nutrition. CKD also increases the risk of cardiovascular diseases. Kidney disease and cardiac failure are linked, and both have common risk factors [30]. As a person's health deteriorates, it can lead to other diseases, such as diabetes or high blood pressure [31]. Kidney transplants and dialysis are mostly needed to keep the kidneys functioning.

16.1.4 Dialysis

There are two forms of dialysis — hemodialysis and peritoneal dialysis, but our study focuses on hemodialysis [32]. The principle of operation of a dialysis machine is such that the blood circuit is the first part of the hemodialysis system. This circuit has an access site, directly on the patient's body at the spot that removes the fluids through the venous tube. In dialysis, the blood flow rate is between 200 and 600 mL per minute [33]. Two needles are connected to the site with a tube, and small amounts of blood are withdrawn from the body through one of the two needles to be pumped into the blood circuit. The blood circuit is connected with the patient through the fistula [34], where a minor surgery is performed to connect an artery and a vein, and it is often done in one of the arms. The graft could also be used directly in place of the artery with a nearby vein. And finally, a temporary or permanent catheter is placed [35].

16.2 Related literature

Iswahyudi et al. [36] in their work designed and implemented a control using electronics where blood is pumped into a hemodialysis machine. The study seeks to create as well as assess the effectiveness of the electronics used in the control of the pump to match the rate of flow in the hemodialysis setup. The fluid was passed through the pump according to its densities. The glucose flow rate was 20% with an accuracy of 0.06% for 5 revolutions per minute and a targeted capacity of 400 mL. Numerous hardware issues, such as roller deterioration and hose elasticity, might reduce the efficacy of a pump system [36]. To a higher standard, the study conducted by Petridis et al. [37] stated the tips used in designing hemodialysis catheters that influence thrombosis as well as the rate of replacement. The study had a 2-year follow-up, prefeasibility research assessed the functionality, and risk of complications of two central venous tunneled catheters (CVTCs) in individuals with end-stage kidney impairment (ESRD). These individuals were administered either a split-tip CVTC or a shotgun-tipped model in the study of ESRD. The patients were all dialyzed within 24 hours. A total of 185 patients were used in the study; 93 were administered with split-tip CVTC, while the remaining 92 received shotgun CVTC. The frequency of CVTC removal was substantially greater in the split tip group, which is 19.3%, than in the shotgun group, which is 8.3% [37].

16.3 Engineering components

16.3.1 Peristaltic pump

Meanwhile, a specialized pump that differs in its design according to the function it performs is used in hemodialysis. Each of these pumps is provided with several controls and safety devices, such as valves and meters for measuring blood pressure outside and inside the patient's body [38]. The blood pump is located before the filter in the blood circuit, and the most common type of pump is the peristaltic pump, a roller pump that works according to the principle of a rolling cylinder [39]. In other words, when the device is turned on and the treatment process begins, the blood pump accelerates the blood flow outside the patient's body to pass through a needle, thereby generating pressure. When the flow increases, the pressure decreases. The dialysis device checks the pressure in the blood tubes before the blood reaches the pump [40]. The pump speed can be adjusted through a switch, which helps to raise the level of blood inside the artery. The pump is equipped with a separate crank handle, which returns the blood to the patient during a power outage. The pump cover is opened, the rotor lock is turned outward, and the arm is turned in a clockwise direction at a rate of 6−10 revolutions per minute, which is equal to 60−100 mL/min [41]. The blood circuit usually includes another pump, the heparin pump, as heparin is an anticoagulant, a chemical used to prevent the natural clotting of the blood, and therefore it is generally used to prolong the time that the blood needs to clot during treatment [42]. A continuous infusion of heparin is the safest way to control the concentration of heparin in the blood. So in most cases, the heparin pump is in the form of a syringe attached to a holder in the dialysis machine, and a piston slowly pumps heparin from it. The syringe is connected to a thin tube that drains at a point where the pressure is positive, i.e., after the blood pump. The pump can adapt to commercially available syringe sizes [42,43]. Nonetheless, the presence of bubbles inside the tubes that carry the (purified) blood entering the patient's body [44,45] poses a great threat. The ultrasonic technique is one of the most widely used sensors for detecting air bubbles, and it is usually placed at the bottom end of the drip chamber [46,47]. On the other hand, the photoelectric technique relies on the photovoltaic system (transmitter and receiver), where light of a certain wavelength is sent through a light-emitting diode (LED) toward the tube that contains the patient's returning blood. The rays from the device are scattered in the form of bubbles [48]. Another important device in hemodialysis is the pressure module for blood located before the drip chamber, which automatically measures the patient's pressure during relaxation. [48,49]. The most important part of the hemodialysis system is the dialyzer, which has the same principle as the kidney in vivo. The dialyzer mediates both circuits to rid the blood of waste products as well as additional fluids. The dynamics of fluids, filtration, diffusion, and osmotic pressure of dialysis are well applied in the dialyzer [50,51]. Dialysis fluid is a mixture of the blood and the dialysis fluid, and it is important to maintain an appropriate level of electrolytes in the patient's blood. The process of removing waste from the body starts in the cells, where solutes (such as waste products and some electrolytes) slowly cross the membranes of the cells and enter the bloodstream [52].

16.3.2 Dialyzers of various types

The hemodialyzer is made up of basically two parts: model design and working principles [53,54]. Also, they can be classified according to the configuration of the dialyzer [50]. They are the flat and hollow-fiber types, which are still in use, whereas the other types are obsolete. The hollow fiber is made of about 1200 hollow fibers with a diameter of 200–300 um as well as 2–30 um in thickness. The dialyzer could also be classified according to the membrane material [55]. This has about four categories: cellulose regenerated membrane made up of copper imitation and ammonia dialyzer membrane. The acetate membrane cellulose dialyzer, synthetic fiber membrane, and replacement fiber membrane dialyzer are all acetate membrane cellulose dialyzers [56]. A dialysis machine is also classified according to the coefficient of ultrafiltration. They are low-coefficient ultrafiltration and high-efficiency dialyzers, which include the blood imitation membrane and cellulose acetate membrane [57]. In addition, the dialyzer could be described as a coil, for it consists of two cellophane tubes, each with a circumference of 9 cm and a length of 10.8 m [58]. It is placed flat on a nylon net and applied in the form of a file [56]. The parallel plate dialyzer is another type of device that consists of epoxy plates in which precise longitudinal channels are made in a layer of cellophane or profane and placed between each of them [59]. The most widely used dialyzer in dialysis is the hollow fiber dialyzer because it is used by 90% of patients, consisting of more than 3000 micro-tubes with a very small diameter of almost 0.001 m and is semipermeable [60,61]. Sterility is highly maintained by daily chemical cleaning and thermal sterilization operations, which are highly effective in the reduction of germs [62].

16.3.3 Materials and methods

The wires and LED indicator are soldered on the breadboard. The ground (GND) and Vin pin headers of the power cable of the Arduino can be used to connect the leads from a battery. The board can run on 6 to 20 V from an external power supply. The four pins are connected to the power supply, trigger, GND, and pump, as can be seen in Fig. 16.2. Water is poured on the 3D plastics and tinted with a dye to differentiate the aspirations during the flow movement. The pressures from the pump allow the pumping of blood and dialysate flows, in and out of the fistula, which was made via a minor surgical scar in an ideal dialyzer machine. The procedure according to our study is simplified in Fig. 16.3. The aim is to pump the water from one of the plastics to the other in a continuous direction. It can also flow in an alternate direction depending on the connections, but in either case, a steady current in milliseconds is applied. This mimics the hemodialyzer machine with an inlet and outlet. The signal goes to the

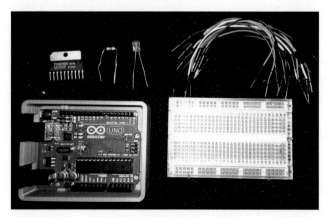

Figure 16.2 Setup of materials: https://www.instructables.com/Control-peristaltic-pump-with-TA7291P-and-an-Ardui/Accessed: 2022-08-22.

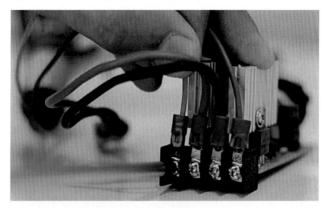

Figure 16.3 Circuit connection of the wires [63].

Figure 16.4 Peristaltic pump and 3D plastics with fluid [64].

Arduino when the circuit is closed. The power source can be from a direct current of 12 V or an alternating current that powers the peristaltic pump. The LED serves as an indicator for the peristaltic pump and is connected parallel to the board. The peristaltic pump squeezes the tube to get the liquid to get the tinted water from one of the 3D plastics to the other and fills the second 3D. This is exactly of the same principle as when the dialyzer removes the blood filters and reintroduces them inside the body. The prototype, as shown in Fig. 16.4, and complete set up Fig. 16.5, is a novel design that could be used in any place, irrespective of geographical settings.

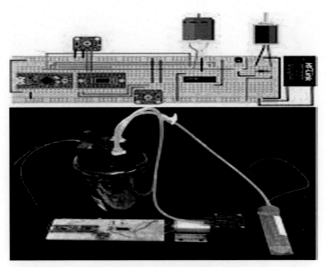

Figure 16.5 Prototype of a hemodialysis circuit [65].

16.4 Conclusion

In our design using Arduino, an electronic device that could be incorporated into different projects and sourced openly was soldered on a breadboard to demonstrate the actions of the peristaltic pump in a continuous flow. The LED indicator, when the circuit is closed, shows that the blood circuit in hemodialysis mimics the functional status of the circulatory pathway and the dialyzer, which work in the same principle as the kidney in vitro. This design maintains the flow rate of 200−600 mL of blood per minute as seen in dialyzers used in health settings. The big-budget procedure is cut down in design, and patients prone to septicemia are reduced drastically. The project, though very significant, cannot be replaced with the conventional dialyzer.

References

[1] S.P. Willig, Kidney anatomy and physiology, Biomedical Instrumentation & Technology / Association for the Advancement of Medical Instrumentation 27 (4) (1993) 342−344. Available from: https://doi.org/10.1007/978-3-319-23458-8_1.
[2] Y.S. Tang, Y.C. Tsai, T.W. Chen, S.Y. Li, Artificial kidney engineering: the development of dialysis membranes for blood purification, Membranes 12 (2) (2022) 177. Available from: https://doi.org/10.3390/MEMBRANES12020177.
[3] P.A. Yusuf, et al., Recent updates in artificial kidney technology: potential parsing for universal coverage burden of chronic kidney disease in Indonesia, AIP Conference

Proceedings 2344 (1) (2021) 050020. Available from: https://doi.org/10.1063/5.0047542.

[4] Anatomy, Abdomen and Pelvis, Kidneys—PubMed. https://pubmed.ncbi.nlm.nih.gov/29494007/ (accessed July 13, 2022).

[5] K. Wager, D. Chari, S. Ho, T. Rees, O. Penner, B.J.A. Schijvenaars, Identifying and validating networks of oncology biomarkers mined from the scientific literature, Cancer Informatics 21 (2022). Available from: https://doi.org/10.1177/11769351221086441.

[6] J.S. Mattoon, R. Pollard, T. Wills, C.R. Berry, Ultrasound-guided aspiration and biopsy procedures, Small Animal Diagnostic Ultrasound (2020) 49−75. Available from: https://doi.org/10.1016/B978-0-323-53337-9.00011-3.

[7] Kidneys: anatomy, Concise Medical Knowledge. https://www.lecturio.com/concepts/kidneys/ (accessed July13, 2022).

[8] Components of the urinary system, SEER Training. https://training.seer.cancer.gov/anatomy/urinary/components/ (accessed July 13, 2022).

[9] A. Chvátal, Discovering the structure of nerve tissue: part 1: from marcello malpighi to christian berres, Journal of the History of the Neurosciences 24 (3) (2015) 268−291. Available from: https://doi.org/10.1080/0964704X.2014.977676.

[10] S. Cohen, H. Wanibuchi, S. Fukushima, Lower urinary tract, Handbook of Toxicologic Pathology (2002) 337−362. Available from: https://doi.org/10.1016/B978-012330215-1/50035-1.

[11] T. Traitteur, C. Zhang, R. Morizane, The application of iPSC-derived kidney organoids and genome editing in kidney disease modeling, iPSCs—State of the Science (2022) 111−136. Available from: https://doi.org/10.1016/B978-0-323-85767-3.00007-4.

[12] 41.11: Human Osmoregulatory and Excretory Systems—Nephron—The Functional Unit of the Kidney—Biology LibreTexts. https://bio.libretexts.org/Bookshelves/Introductory_and_General_Biology/Book%3A_General_Biology_(Boundless)/41%3A_Osmotic_Regulation_and_the_Excretory_System/41.11%3A__Human_Osmoregulatory_and_Excretory_Systems_-_Nephron-_The_Functional_Unit_of_the_Kidney (accessed July 13, 2022).

[13] S. De, R. Nishinakamura, Nephron progenitors in induced pluripotent stem cell−derived kidney organoids, iPSC Derived Progenitors (2022) 201−213. Available from: https://doi.org/10.1016/B978-0-323-85545-7.00013-2.

[14] S. Aymé, et al., Common elements in rare kidney diseases: conclusions from a kidney disease: improving global outcomes (KDIGO) controversies conference, Kidney International 92 (4) (2017) 796−808. Available from: https://doi.org/10.1016/J.KINT.2017.06.018.

[15] P. Rajkumar, J.L. Pluznick, Acid-base regulation in the renal proximal tubules: using novel pH sensors to maintain homeostasis, American Journal of Physiology - Renal Physiology 315 (5) (2018) F1187. Available from: https://doi.org/10.1152/AJPRENAL.00185.2018.

[16] Physiology, renin angiotensin system—statpearls—NCBI bookshelf. https://www.ncbi.nlm.nih.gov/books/NBK470410/ (accessed July 13, 2022).

[17] The 7 functions of the kidneys, FKP Kidney Doctors. https://flkidney.com/the-7-functions-of-the-kidneys/ (accessed June 23, 2022).

[18] V. Jha, et al., Chronic kidney disease: global dimension and perspectives, Lancet 382 (9888) (2013) 260−272. Available from: https://doi.org/10.1016/S0140-6736(13)60687-X.

[19] Kidney Disease Statistics for the United States, NIDDK. https://www.niddk.nih.gov/health-information/health-statistics/kidney-disease. (accessed August 22, 2022).

[20] The Kidney Project UCSF. https://pharm.ucsf.edu/kidney/need/statistics. (accessed August 22, 2022).

[21] M.S. Matovinović, Pathophysiology and classification of kidney disease 1. Pathophysiology and Classification of Kidney Diseases. Available: http://www.ifcc.org.

[22] K.J. Jager, C. Kovesdy, R. Langham, M. Rosenberg, V. Jha, C. Zoccali, A single number for advocacy and communication-worldwide more than 850 million individuals have kidney diseases, Kidney International 96 (5) (2019) 1048–1050. Available from: https://doi.org/10.1016/J.KINT.2019.07.012.

[23] C.M. Rhee, C.P. Kovesdy, Spotlight on CKD deaths—increasing mortality worldwide, Nature Reviews Nephrology 11 (4) (2015) 199. Available from: https://doi.org/10.1038/NRNEPH.2015.25.

[24] Global facts: about kidney disease. https://www.kidney.org/kidneydisease/global-facts-about-kidney-disease (accessed July 13, 2022).

[25] B. Bikbov, et al., Global, regional, and national burden of chronic kidney disease, 1990–2017: a systematic analysis for the global burden of disease study 2017, The Lancet 395 (10225) (2020) 709–733. Available from: https://doi.org/10.1016/S0140-6736(20)30045-3/ATTACHMENT/234A6931-5886-48BE-8997-33DC6E6B8DBB/MMC1.PDF.

[26] Chronic Kidney Disease, 2024. Available from: https://www.worldkidneyday.org/facts/chronic-kidney-disease/.

[27] Diseases of the kidney and the urinary system—disease control priorities in developing countries, NCBI Bookshelf. https://www.ncbi.nlm.nih.gov/books/NBK11791/ (accessed June 24, 2022).

[28] Kidney disease: the basics. https://www.kidney.org/news/newsroom/fsindex (accessed July 13, 2022).

[29] A.L. Ammirati, Chronic kidney disease, Revista da Associacao Medica Brasileira 66 (2020) 3–9. Available from: https://doi.org/10.1590/1806-9282.66.S1.3. Associacao Medica Brasileira.

[30] F.T. Chebib, V.E. Torres, Recent advances in the management of autosomal dominant polycystic kidney disease, Clinical Journal of the American Society of Nephrology 13 (11) (2018) 1765–1776. Available from: https://doi.org/10.2215/CJN.03960318.

[31] Annual Data Report. https://adr.usrds.org/2020/ (accessed June 23, 2022).

[32] Annual Data Report. https://www.usrds.org/annual-data-report/ (accessed June. 23, 2022).

[33] J.Y. Yeun, T.A. Depner, S. Ananthakrishnan, Principles of hemodialysis, Chronic Kidney Disease, Dialysis, and Transplantation: A Companion to Brenner and Rector's the Kidney, 2018, pp. 339–360.e7. Available from: https://doi.org/10.1016/B978-0-323-52978-5.00022-7.

[34] E. Balikci, B. Yilmaz, A. Tahmasebifar, E.T. Baran, E. Kara, Surface modification strategies for hemodialysis catheters to prevent catheter-related infections: a review, Journal of Biomedical Materials Research, Part B: Applied Biomaterials 109 (3) (2021) 314–327. Available from: https://doi.org/10.1002/JBM.B.34701.

[35] Y. Pradeep Vaidya, G. Randall Green, D. Chief, C. Yash Pradeep Vaidya, Coronary artery fistula, Journal of Cardiac Surgery 34 (2) (2019) 1608–1616. Available from: https://doi.org/10.1111/JOCS.14267.

[36] A. Iswahyudi, P. Busono, Y. Suryana, A. Mujadin, D. Astharini, Design and implementation of electronic control system of blood pump for hemodialysis machine, Jurnal Al-Azhar Indonesia Seri Sains Dan Teknologi 3 (3) (2017) 150. Available from: https://doi.org/10.36722/SST.V3I3.221.

[37] C. Petridis, et al., Tip design of hemodialysis catheters influences thrombotic events and replacement rate, European Journal of Vascular and Endovascular Surgery. The Official Journal of the European Society for Vascular Surgery 53 (2) (2017) 262–267. Available from: https://doi.org/10.1016/J.EJVS.2016.10.015.

[38] H. Kawanishi, M. Moriishi, N. Takahashi, S. Tsuchiya, The central dialysis fluid delivery system (CDDS): is it specialty in Japan? Renal Replacement Therapy 2 (1) (2016) 1−8. Available from: https://doi.org/10.1186/S41100-016-0016-4/FIGURES/4.

[39] P.A. Smith, Y. Wang, S. Gro-Hardt, R. Graefe, Hydraulic design, Mechanical Circulatory and Respiratory Support, 2018, pp. 301−334. Available from: https://doi.org/10.1016/B978-0-12-810491-0.00010-2.

[40] Physiology, airflow resistance − StatPearls, NCBI Bookshelf. https://www.ncbi.nlm.nih.gov/books/NBK554401/ (accessed July 13, 2022).

[41] D.R.M. Hassell, F.M. van der Sande, J.P. Kooman, J.P. Tordoir, K.M.L. Leunissen, Optimizing dialysis dose by increasing blood flow rate in patients with reduced vascular-access flow rate, American Journal of Kidney Diseases 38 (5) (2001) 948−955. Available from: https://doi.org/10.1053/AJKD.2001.28580.

[42] H. Hebibi, et al., Arterial versus venous port site administration of nadroparin for preventing thrombosis of extracorporeal blood circuits in patients receiving hemodiafiltration treatment, Kidney International Reports 6 (2) (2021) 351−356. Available from: https://doi.org/10.1016/J.EKIR.2020.11.020.

[43] M. Murea, et al., Efficacy and safety of low-dose heparin in hemodialysis, Hemodialysis International 22 (1) (2018) 74−81. Available from: https://doi.org/10.1111/HDI.12563.

[44] (PDF) Air bubbles detection and alarm in the blood stream of dialysis using capacitive sensors. https://www.researchgate.net/publication/330838362_Air_bubbles_detection_and_alarm_in_the_blood_stream_of_dialysis_using_capacitive_sensors (accessed June 24, 2022).

[45] (PDF) Capacitive air bubble detector operated at different frequencies for application in hemodialysis. https://www.researchgate.net/publication/242602131_Capacitive_Air_Bubble_Detector_Operated_at_Different_Frequencies_for_Application_in_Hemodialysis (accessed June 24, 2022).

[46] H. Wang, Z. Chen, Y. Chen, M. Xie, L. Hua, Mechanism study of bubble removal in narrow viscous fluid by using ultrasonic vibration, Japanese Journal of Applied Physics 58 (11) (2019). Available from: https://doi.org/10.7567/1347-4065/AB4DFA.

[47] Y. Li, J. Wu, L. Fu, J. Wang, A fast bubble detection method in microtubes based on pulsed ultrasound, Micromachines (Basel) 12 (11) (2021) 1−8. Available from: https://doi.org/10.3390/mi12111402.

[48] J.X. Wu, P.T. Huang, C.H. Lin, C.M. Li, Blood leakage detection during dialysis therapy based on fog computing with array photocell sensors and heteroassociative memory model, Healthcare Technology Letters 5 (1) (2018) 38. Available from: https://doi.org/10.1049/HTL.2017.0091.

[49] K. Matsuda, R. Fissell, S. Ash, B. Stegmayr, Long-term survival for hemodialysis patients differ in Japan versus Europe and the USA. What might the reasons be? Artificial Organs 42 (12) (2018) 1112−1118. Available from: https://doi.org/10.1111/AOR.13363.

[50] M. Abe, I. Masakane, A. Wada, S. Nakai, K. Nitta, H. Nakamoto, Dialyzer classification and mortality in hemodialysis patients: a 3-year nationwide cohort study, Frontiers in Medicine 8 (2021) 1468. Available from: https://doi.org/10.3389/FMED.2021.740461/BIBTEX.

[51] IHow does dialysis work?—InformedHealth.org—NCBI Bookshelf. https://www.ncbi.nlm.nih.gov/books/NBK492981/ (accessed June 24, 2022).

[52] O. Swift, E. Vilar, K. Farrington, Hemodialysis, Medicine (United Kingdom) 47 (9) (2022) 596−602. Available from: https://doi.org/10.1016/j.mpmed.2019.06.004.

[53] A.T. Azar, B. Canaud, Hemodialysis system, Studies in Computational Intelligence 404 (2013) 99−166. Available from: https://doi.org/10.1007/978-3-642-27458-9_3.

[54] (PDF) L4 Hemodialysis Machine. https://www.researchgate.net/publication/325662967_L4_Hemodialysis_Machine (accessed July 13, 2022).

[55] M. Abe, T. Hamano, A. Wada, S. Nakai, I. Masakane, Effect of dialyzer membrane materials on survival in chronic hemodialysis patients: results from the annual survey of the Japanese nationwide dialysis registry, PLoS One 12 (9) (2017). Available from: https://doi.org/10.1371/JOURNAL.PONE.0184424.

[56] Y.A. Chen, S.M. Ou, C.C. Lin, Influence of dialysis membranes on clinical outcomes: from history to innovation, Membranes 12 (2) (2022). Available from: https://doi.org/10.3390/membranes12020152. MDPI.

[57] What Are the Types of Dialyzers?—Lepu Medical Technology (Beijing) Co., Ltd. https://en.lepumedical.com/what-are-the-types-of-dialyzers.html (accessed June 24, 2022).

[58] B.W. Teo, N.S. Kanagasundaram, E.P. Paganini, Continuous renal replacement therapy, Critical Care Medicine (2008) 301−326. Available from: https://doi.org/10.1016/B978-032304841-5.50021-2.

[59] M. Kahshan, D. Lu, M. Rahimi-Gorji, Hydrodynamical study of flow in a permeable channel: application to flat plate dialyzer, International Journal of Hydrogen Energy 44 (31) (2019) 17041−17047. Available from: https://doi.org/10.1016/J.IJHYDENE.2019.04.211.

[60] P. Susantitaphong, B.L. Jaber, Methods and complications of dialyzer reuse, Handbook of Dialysis Therapy, fifth ed., Elsevier, 2017, pp. 144−151.e1. doi: 10.1016/B978-0-323-39154-2.00011-4.

[61] G. Eknoyan, et al., Effect of dialysis dose and membrane flux in maintenance hemodialysis, New England Journal of Medicine 347 (25) (2002) 2010−2019. Available from: https://doi.org/10.1056/nejmoa021583.

[62] W.A. Rutala, D.J. Weber, Disinfection, sterilization, and control of hospital waste, Mandell, Douglas, and Bennett's Principles and Practice of Infectious Diseases 2 (2015) 3294. Available from: https://doi.org/10.1016/B978-1-4557-4801-3.00301-5.

[63] Circuit Construction, 2018. https://learn.adafruit.com/reef-pi-guide-5-dosing-controller/circuit-construction.

[64] 3D Printed Peristaltic Pump Has Impressive Capabilities, 2023. https://hackaday.com/2015/11/10/3d-printed-peristaltic-pump-has-impressive-capabilities/.

[65] What is hemodialysis? 2022. https://kidneycampus.ca/what-is-hemodialysis/.

CHAPTER SEVENTEEN

Design of endoscopic medical device

Dilber Uzun Ozsahin[1,2,3]**, Declan Ikechukwu Emegano**[3,4]**,**
Abdulsamad Hassan[4]**, Mohammad Aldakhil**[4]**, Ali Mohsen Banat**[4]**,**
Basil Bartholomew Duwa[3,4] **and Ilker Ozsahin**[3,5]

[1]Department of Medical Diagnostic Imaging, College of Health Science, University of Sharjah, Sharjah, United Arab Emirates
[2]Research Institute for Medical and Health Sciences, University of Sharjah, Sharjah, United Arab Emirates
[3]Operational Research Center in Healthcare, Near East University, Nicosia/TRNC, Mersin 10, Turkey
[4]Department of Biomedical Engineering, Near East University, Nicosia/TRNC, Mersin 10, Turkey
[5]Department of Radiology, Brain Health Imaging Institute, Weill Cornell Medicine, New York, NY, United States

Contents

17.1 Introduction

An endoscope is a medical device widely used in vivo for imaging as well as for clinical evaluations of cancers, especially of the digestive tract [1]. This instrument examines all the internal organs when inserted through natural openings such as the mouth and anus. The instrument has come a long way in modern times, and it can now be used in both optics and mechanics. The medical endoscope is embedded with a camera, and its functions cannot be understood without full comprehension of endoscopic functions [2]. The discovery of the endoscope as a medical tool has changed the medical approach to diagnoses and the utilization of

Practical Design and Applications of Medical Devices.
DOI: https://doi.org/10.1016/B978-0-443-14133-1.00013-6
231

a piece of equipment. The new idea and technology have had a huge impact on the medical world [1−3] The process whereby clinicians use an endoscope to view internal organs is termed endoscopy, and it should not be confused with surgery [4,5]. The invention of endoscopy, as a methodology for studying the inside of the human body, has changed the way doctors look at images and their accuracy [6]. Medical endoscopy has been technologically advanced to include different kinds of endoscopies for different parts of the body [7]. Over centuries medical endoscopes have played the role of a user-friendly, flexible, long tubular wire used in internal organs [8].

17.1.1 Historical perspectives on endoscopes

Endoscopy has been used as a medical procedure for a long time. It was discovered in the 1800s. Phillip Bozzini, who made the Lechleiter in 1805, is thought to have made the first endoscope [9,10]. Bee wax provided light for the instrument, which had a silver mirror that reflected the picture. However, the first endoscopes did not have a camera lens. In 1853, Antoine Jean Desormeaux made an endoscope, which was a tool used to look at the urinary system. Since that day, the term "endoscope" has been used to describe the tool [9,11,12]. Improvements in art, architecture, and innovation took years to reach the point where optical viewfinders could be put on endoscopes. The endoscope is what it is today because of the contributions of several health providers over the years. Electronic video–supported endoscopy came into existence in 1983. Welch Alleyn used an electronic sensor for the first time to send an image from the inside of the machine [11]. The image was sent to a monitor by connecting an electronic sensor. At the same time, Olympus Corporation made several improved versions of the electronic medical endoscope, which were widely used for endoscopy. Similarly, Japanese companies also started making electronic endoscopes that were competitive. Fujinon made the first flexible hysteroscope, which came out in 1985 [13]. Since it was first used as a rigid device, there have been modifications to the endoscope. Nowadays, most medical facilities use typical modern tools. With new technology, WiFi endoscopes are now the newest medical discovery [11].

17.1.2 Components of a typical modern endoscope

A typical modern endoscope comprises a lengthy, thin, as well as flexible tube. The image is taken by a lens or camera at the end of one side of the tube. There is an eyepiece and a light system built into the tube [14,15]. The endoscope has a control system and a water tube for cleaning the lens when it has blurred images. The endoscope also has an umbilical cord attached to the processor [16] providing light, and it has hookups to air the suction water. The insertion tool goes inside the patient's body. This is a complex structure with several walls with a polymer coat, a base layer, stainless steel wire mesh, an outer spiral metal band, and an inner spiral metal band [17]. The inner section has an air–water channel, water jet, lens, and object to the lens as well as light guards [18]. When the umbilical cord is hooked to the patient, the light sender and light processor are delivered through the umbilical cord. The air–water button, when pressed, connects to air and water through the light connections. The water cleans the lens in vivo. The cable is used to plug into the electrical system. In addition to these, an endoscope comprises the following parts, using Olympus as a standard [11,19].

17.1.3 Classification of endoscopes

Modern endoscopy systems have advanced diagnostic procedures in terms of image quality, clarity, and treatment given to the patient [19]. The introduction of high-definition (HD) television sparked the emergence of high–resolution images [20]. The best HD endoscopy systems on the market right now have a resolution of 1400 x 1080 pixels [21]. HD endoscopy is used with traditional or virtual chromoendoscopy [22,23]. Chromoendoscopy makes it easier to find colorectal polyps [21,23,24]. Endoscopy is also becoming popular for virtual coloring [19,25]. In autofluorescence imaging with fluorescence intensity image analysis, short light waves are used to ignite the tissue; the energy radioactivity causes the tissue's fluorophores to give off longer wavelengths of visible light [17]. Confocal laser endomicroscopy (CLE) is a histological technique used to view specific areas in the mucosa. CLE systems are of two types: eCLE (endoscope-based CLE) and pCLE (probe-based CLE) [26,27]. Over the past 10 years, wireless capsule endoscopy (WCE) has established itself as the gold standard for the diagnosis of suspected diseases of the small intestine. These include obscure GI bleeding, angiodysplasia, Crohn's disease, celiac disease, polyposis, and tumors [28,29]. WCE entails the ingestion of a miniature pill-sized camera, and once the patient has swallowed the capsule, normal activities are resumed [30,31]. Another classification of endoscopes is based on their rigidity and flexibility. These are long and thin, which makes them easier for medics to use [32,33]. Most of the time, HDF cameras are used in an endoscope, which gives detailed in vivo analysis of the inside of the patient's body [34].

17.2 Journal reviews

The study performed by Katie et al. demonstrates that as a result of the awkward positions, increased effectiveness, and repeated motions required during endoscopic procedures, gastroenterologists often have muscle fatigue and discomfort in their hands and fingers, wrists, forearms, shoulders, and backs. The manufacturers of scopes should be aware of the ergonomic issues that have been raised by the modern design of endoscopes, and they should integrate ergonomic concepts into their future product designs. To reduce the likelihood of experiencing an injury caused by an endoscope, medical professionals need to get training in ergonomic concepts. We take a look at some of the potential developments that may be made in the future to enhance ergonomics and go over some of the necessary adjustments that need to be made to the design of endoscopes so that there is less of a chance of being hurt during endoscopic treatments [35]. According to Omidbakhsh et al. [36], within a system that uses vaporized hydrogen peroxide for sterilization, an endoscope sterilization cycle was designed. This cycle had been put through its paces to research the sterilization of flexible gastrointestinal endoscopes, such as colonoscopes, and duodenoscopes, as well as the interoperability of their components with both authentic flexible gastrointestinal endoscopes and experimentally reconfigured endoscopes that make use of composite materials [36]. Getting mastery of the method of endoscope calls for a significant amount of experience. In their study, Misra et al. [37] developed an add-on robot module that enhances the user experience of conventional endoscopes, making it possible for a single physician to effortlessly operate the device as described. We discovered essential user elements of conventional endoscopes that need to be replicated in a robot's configuration. These features need to be duplicated in conventional endoscopes. In our model, the doctor operates the light driving system with the help of a remote control linked to the system, which makes it possible to control the fully automated endoscope while it is in space. A test was carried out so that we could evaluate how usable our system is. According to the findings, utilizing robotic steering with a position-controlled touchpad boosts both productivity and the level of happiness experienced [37]. In all the reviews, the detailed parts of the essential flexible endoscope were limited. Our study designed a full comprehensive endoscope with internal dimensions of the parts. The design is for knowledge acquisition and cannot replace the endoscopes that are in use.

17.3 Methodology and design of endoscopy devices

The endoscopy imaging device comprises a thermostat with control fans and simulation models. The configuration is made possible by the flexible endoscope's one-of-a-kind design. The GrabCAD Library, which provides millions of completely free computer-aided designs (CADs), and CAD files, including 3D models, is the source of the data that has been provided here. A CAD model of the recursive feature elimination system also includes a flexible endoscopic component. The design of the flexible endoscopic module is divided into four separate elements: a camera module, a flexible forearm, a carbon-fiber shaft, and a mounting backside. Each of these parts serves a specific function in the overall assembly. As can be seen in Fig. 17.1, this module offers three movements: one rolling movement and two orthogonally bending movements. The continuous mechanism, which can be seen in Fig. 17.2A, serves as the foundation for the construction of the flexible wrist part of the endoscope. This mechanism comprises eight disks. The flexible wrist is 26 millimeters in length, and its overall size is 7.5 millimeters. As can be seen in Fig. 17.2B, the camera is a miniature module manufactured by MISUMI Electronic Corporation (model number MDV21106L-128). The camera has a screen resolution of 640 by 480 pixels. The shaft is made of carbon fiber of 416 millimeters in length [38,39]. The fan, as shown in Fig. 17.3, serves as a cooling device, as shown in Figs. 17.1 and 17.4.

Light-emitting diode CRI > 90 5700 K endoscope is the most recent endoscopic device. The source of the light module has a high LED control keypad and an LED screen that alerts users to the presence of a power surge. This lead, in conjunction with the heat sink, constitutes a positive point. However, the body of the box itself may be negative since it is

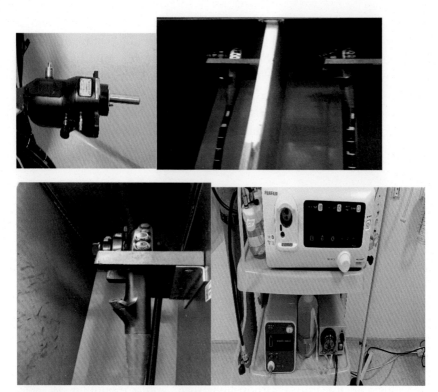

Figure 17.1 Components of the endoscope.

Figure 17.2 LED light source.

isolated by an insulator and has a heatsink and boxes on the ground that are distinct, as shown in Fig. 17.2. This has an Arduino, a breadboard, a connection, and a power supply. The most recent device is an LED endoscopic light source that has a controller and is touch-sensitive. The

Figure 17.3 Light module of the endoscope.

Figure 17.4 The fan.

evolution of LEDs has reached its most recent stage with this innovation. Endoscopy, microscopy, and other diagnostic procedures rely on it. It has a high luminosity, similar to a 350-W source of light, and is completely controlled via remotes. It has a longer life, which may last up to 60000 hours, 120 times longer than xenon, with excellent color and temperatures. It reduces energy consumption by more than half. It is non-harmful to the surrounding ecosystem, user-friendly, and simple in its operations

The cooling system consists of 12 V Noctua fans with less noise. Its performance characteristics are outstanding and of high quality [40] The fan is 12 V with 32 watts of electricity. The heating system is placed directly under the energy efficiency directive. The energy heat transfer takes place by convention. The Raspberry Pi pins maximally provide a driving voltage of 5 V while the Noctua fan is coupled to a DC-DC converter so that the 5 V will be stepped up. During connection, the board is

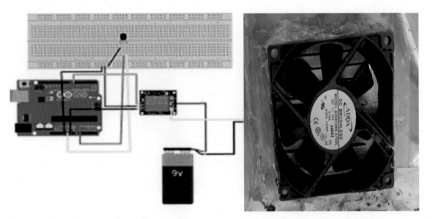

Figure 17.5 The complete circuit.

connected to a Raspberry Pi only. There is also a thermostat, which is connected to the Pi; but in the connections, sensors are used, especially the analog of the Raspberry Pi. Analog to digital connections with a 10- or less-bit channel are of lesser cost and require no additional components. The thermostat takes control of the temperature, whereas the fan speed is controlled depending on the temperature of the system. The connection between the fan and the thermostat is shown in Fig. 17.5.

The speed of the fan is controlled by the temperature of the Raspberry Pi. The Raspberry Pi is efficient and comes with a lot of user-friendly inputs, such as WiFi or Bluetooth, noise filters, and high-speed random access memory. The brightness of the LED could be controlled by the Raspberry Pi at different duty cycles, which either increases the brightness or reduces it, as can be seen in Fig. 17.5.

17.4 Conclusion

The endoscopy imaging device is capable of flexible GI endoscopes. It is sterilized using vaporized hydrogen peroxide. The sterilization techniques of its components are modified to be compatible. Flexible colonoscopy (endoscope) is a dependable, sparsely invasive procedure used to make diagnoses and treat a wide range of gastrointestinal disorders such as stomach aches, heartburn, acid, gastroesophageal reflux disease, lesions, gastric ulcers, swallowing difficulties, digestive tract hemorrhaging, alterations in bowel habits, and adenomas. Before undertaking an endoscopy, providing informational brochures as support for permission is important. Specifically in industrialized nations such as Canada, the United States, and Europe. governments and organizations are implementing stringent regulatory criteria to assist in the development of medical endoscopic equipment that is both safe and efficient. However, this design is not intended for business use and hence cannot be duplicated. It has been developed primarily for academic studies.

References

[1] S. Ahmed, P.R. Galle, H. Neumann, Molecular endoscopic imaging: the future is bright, Therapeutic Advances in Gastrointestinal Endoscopy 12 (2019) 1−15. Available from: https://doi.org/10.1177/2631774519867175.

[2] M.S.U. Osagie, O. Enagbonma, A.I. Inyang, Structural dynamics and evolution of capsule endoscopy (Pill camera) technology in gastroenterologists assertion, International Journal in Foundations of Computer Science & Technology 8 (1/2) (2018) 01−12. Available from: https://doi.org/10.5121/ijfcst.2018.8201.

[3] J.K. Bae, et al., Smartphone-based endoscope system for advanced point-of-care diagnostics: feasibility study, JMIR Mhealth Uhealth 5 (7) (2017). Available from: https://doi.org/10.2196/MHEALTH.7232.

[4] N.S. Grenager, J.A. Orsini, Endoscopy, Comparative Veterinary Anatomy (2022) 18−22. Available from: https://doi.org/10.1016/B978-0-323-91015-6.00002-9. Jan.

[5] E.J. Kezirian, R.J. Schwab, Upper airway imaging and endoscopy, Encyclopedia of Sleep (2013) 479−489. Available from: https://doi.org/10.1016/B978-0-12-378610-4.00354-5. Jan.

[6] B. Münzer, K. Schoeffmann, L. Böszörmenyi, Content-based processing and analysis of endoscopic images and videos: a survey, Multimedia Tools and Applications 77 (1) (2017) 1323−1362. Available from: https://doi.org/10.1007/S11042-016-4219-Z. Jan.

[7] Endoscopy - Better Health Channel. https://www.betterhealth.vic.gov.au/health/conditionsandtreatments/endoscopy (accessed Jun. 29, 2022).

[8] Y. Moon, et al., Cost-effective smartphone-based articulable endoscope systems for developing countries: instrument validation study, JMIR Mhealth Uhealth 8 (9) (2020). Available from: https://doi.org/10.2196/17057.

[9] D. Ramai, K. Zakhia, D. Etienne, M. Reddy, Philipp Bozzini (1773−1809): The earliest description of endoscopy, Journal of Medical Biography 26 (2) (2018) 137−141. Available from: https://doi.org/10.1177/0967772018755587.

[10] S.H. Lee, Y.K. Park, S.M. Cho, J.K. Kang, D.J. Lee, Technical skills and training of upper gastrointestinal endoscopy for new beginners, World Journal of Gastroenterology: WJG 21 (3) (2015) 759. Available from: https://doi.org/10.3748/WJG.V21.I3.759.

[11] Who developed the endoscope?, United Endoscopy. https://www.endoscope.com/blog/who-developed-the-endoscope/ (accessed Jun. 29, 2022).

[12] Origin of Endoscopes, Endoscopes, History of Olympus Products, Technology, Olympus. https://www.olympus-global.com/technology/museum/endo/?page = technology_museum (accessed Jun. 29, 2022).

[13] Olympus Launches ENDO-AID, an AI-Powered Platform for Its Endoscopy System - Olympus EMEA., Medical Systems, 2020. https://www.olympus-europa.com/company/en/news/press-releases/2020-10-09t08-30-00/olympus-launches-endo-aid-an-ai-powered-platform-for-its-endoscopy-system.html (accessed Jul. 27, 2022).

[14] N. Kurniawan, M. Keuchel, Flexible gastro-intestinal endoscopy—clinical challenges and technical achievements, Comput Struct Biotechnol Journal 15 (2017) 168. Available from: https://doi.org/10.1016/J.CSBJ.2017.01.004.

[15] Types of endoscopy, Cancer.Net. https://www.cancer.net/navigating-cancer-care/diagnosing-cancer/tests-and-procedures/types-endoscopy (accessed Aug. 26, 2022).

[16] "Outline of Gastrointestinal Endoscopy System."

[17] ASGE, Endoscopic Procedures. https://www.asge.org/home/about-asge/newsroom/media-backgrounders-detail/endoscopic-procedures (accessed Aug. 26, 2022).

[18] D. Kreeft, E.A. Arkenbout, P.W.J. Henselmans, W.R. van Furth, P. Breedveld, Review of techniques to achieve optical surface cleanliness and their potential application to surgical endoscopes, Surgical Innovation 24 (5) (2017) 509. Available from: https://doi.org/10.1177/1553350617708959.

[19] Types of endoscopy, Cancer.Net. https://www.cancer.net/navigating-cancer-care/diagnosing-cancer/tests-and-procedures/types-endoscopy (accessed Jun. 29, 2022).

[20] Robinson, J., Holland, S., Runco, S., Pitts, D., Whitehead, V., & Andrefouet, S. (2000). High-Definition Television (HDTV) Images for Earth Observations and Earth Science Applications. https://www.researchgate.net/publication/24314832_High-Definition_Television_HDTV_Images_for_Earth_Observations_and_Earth_Science_Applications (accessed Jul. 27, 2022).

[21] J.W. Rey, New aspects of modern endoscopy, World Journal of Gastrointest Endoscopy 6 (8) (2014) 334. Available from: https://doi.org/10.4253/WJGE.V6.I8.334.

[22] V. Subramanian, et al., Comparison of high definition with standard white light endoscopy for detection of dysplastic lesions during surveillance colonoscopy in patients with colonic inflammatory bowel disease, Inflammatory Bowel Diseases 19 (2) (2013) 350–355. Available from: https://doi.org/10.1002/IBD.23002.

[23] R. Singh, K.H. Chiam, F. Leiria, L.Z.C.T. Pu, K.C. Choi, M. Militz, Chromoendoscopy: role in modern endoscopic imaging, Translational Gastroenterology and Hepatology 5 (2020). Available from: https://doi.org/10.21037/TGH.2019.12.06. Jul.

[24] P.J. Trivedi, B. Braden, Indications, stains and techniques in chromoendoscopy, QJM: An International Journal of Medicine 106 (2) (2013) 117. Available from: https://doi.org/10.1093/QJMED/HCS186.

[25] S.F. Pasha, et al., Comparison of the yield and miss rate of narrow band imaging and white light endoscopy in patients undergoing screening or surveillance colonoscopy: a meta-analysis, The American Journal of Gastroenterology 107 (3) (2012) 363–370. Available from: https://doi.org/10.1038/AJG.2011.436.

[26] K. Karia, M. Kahaleh, A review of probe-based confocal laser endomicroscopy for pancreaticobiliary disease, Clinical Endoscopy 49 (5) (2016) 462. Available from: https://doi.org/10.5946/CE.2016.086.

[27] A. Ciocâlteu, et al., Evaluation of new morphometric parameters of neoangiogenesis in human colorectal cancer using confocal laser endomicroscopy (CLE) and targeted pan endothelial markers, PLoS One 9 (3) (2014) e91084. Available from: https://doi.org/10.1371/JOURNAL.PONE.0091084.

[28] A.G. Ionescu, et al., Clinical impact of wireless capsule endoscopy for small bowel investigation (review, Experimental and Therapeutic Medicine 23 (4) (2022) 1–9. Available from: https://doi.org/10.3892/ETM.2022.11188.

[29] V.V. Zhirnov, R.K. Cavin, Microsystems for bioelectronics, Microsystems for Bioelectronics (2011). Available from: https://doi.org/10.1016/C2009-0-20467-3.

[30] About capsule endoscopy - capsovision international. https://capsovision.com/international/patient-resources/about-capsule-endoscopy/ (accessed Jul. 27, 2022).

[31] R. Kirthi, A. Krishna Sailaja, Medical capsule robots: an overview on wireless capsule endoscope, Pharmaceutics and Pharmacology Research 5 (4) (2022) 01–04. Available from: https://doi.org/10.31579/2693-7247/072.

[32] Endoscopy classification - ultrasonic spray coating - cheersonic. https://cheersonic-liquid.com/en/endoscopy-classification/ (accessed Jul. 27, 2022).

[33] J. Hochberger, V. Meves, G.G. Ginsberg, Difficult cannulation and sphincterotomy, Clinical Gastrointestinal Endoscopy (2019) 563−570. Available from: https://doi.org/10.1016/B978-0-323-41509-5.00050-5. e2.

[34] Noureldin, Y. & Andonian, S. (2016). Laparoscopic/robotic camera and lens systems, Abdominal Key. https://abdominalkey.com/laparoscopicrobotic-camera-and-lens-systems/ (accessed Jul. 27, 2022).

[35] K. Schwab, S. Singh, An introduction to flexible endoscopy, Surgery (Oxford) 29 (2) (2011) 80−84. Available from: https://doi.org/10.1016/J.MPSUR.2010.11.014.

[36] N. Omidbakhsh, S. Manohar, R. Vu, K. Nowruzi, Flexible gastrointestinal endoscope processing challenges, current issues, and future perspectives, Journal of Hospital Infection 110 (2021) 133−138. Available from: https://doi.org/10.1016/J.JHIN.2021.01.021. Apr.

[37] S.P. Misra, M. Dwivedi, K. Sharma, Colon tumors and colonoscopy, Endoscopy 43 (11) (2011) 985−989, doi:10.1055/s-0031-1291436.

[38] Z. Li and R. Du, Design and analysis of a bio-inspired wire-driven multi-section flexible robot. *International Journal of Advanced Robotic Systems.* 10, pp. 209-220, 2013, doi: 10.5772/56025.

[39] C. Song, X. Ma, X. Xia, P.W.Y. Chiu, C.C.N. Chong, Z. Li, A robotic flexible endoscope with shared autonomy: a study of mockup cholecystectomy, Surgical Endoscopy 34 (6) (2020) 2730−2741. Available from: https://doi.org/10.1007/s00464-019-07241-8.

[40] Fans. https://noctua.at/en/products/fan (accessed Jun. 29, 2022).

CHAPTER EIGHTEEN

Temperature reducer machine

Dilber Uzun Ozsahin[1,2,3], Declan Ikechukwu Emegano[3,4], Riad Alsabbagh[4], Basil Bartholomew Duwa[3,4] and Ilker Ozsahin[3,5]

[1]Department of Medical Diagnostic Imaging, College of Health Science, University of Sharjah, Sharjah, United Arab Emirates
[2]Research Institute for Medical and Health Sciences, University of Sharjah, Sharjah, United Arab Emirates
[3]Operational Research Center in Healthcare, Near East University, Nicosia/TRNC, Mersin 10, Turkey
[4]Department of Biomedical Engineering, Near East University, Nicosia/TRNC, Mersin 10, Turkey
[5]Department of Radiology, Brain Health Imaging Institute, Weill Cornell Medicine, New York, NY, United States

Contents

18.1 Introduction

The temperature of the deep body tissues, which make up the core of the skin, stays stable, within $1°F$ ($0.6°C$), unless a person has a feverish illness. A naked individual subjected to temperatures ranging from $55°F$ to as high as $130°F$ in warmer air may retain a core body temperature that is almost consistent [1]. In comparison to the actual human temperature, the epidermis (skin) temperature value fluctuates based on environmental and atmospheric conditions. When discussing the skin's capacity to transfer heat to its environment, the natural skin temperature is crucial [2]. Fever occurs when the core body temperature of an individual rises above the set point, which is regulated by the thermoregulators of the hypothalamus [3]. The rise in body temperature at a set point is secondary compared with the increase as a result of immunological responses. In humans, the normal body temperature is $37°C$, though it varies throughout 24 hours. This variation

Practical Design and Applications of Medical Devices.
DOI: https://doi.org/10.1016/B978-0-443-14133-1.00005-7

though minimal is a result of physiological changes in the body, which is normally about 0.5°C. However, when the core body temperature increases above 0.5°C as a result of fever, it is due to unknown origin pyrogens [4,5]. Fever is the main way that both warm-blooded and cold-blooded organisms react to an infectious disease, and it has been like this for over 600 million years. Physiological, as well as neuronal circuitry, work together to make the fever response happen, and it helps the body fight off an infectious disease [6]. Celsius, who lived in the first century BC, said that there are four main signs of the inflammatory response: pain (dolor), redness (rubor), swelling (tumor), and fever. There seems to be more proof now that the 1- to 4-°C rise in core temperature that happens during a fever helps people live longer and get rid of many infectious diseases [7,8]. Using antipyretic prescription medications to reduce fever is linked to a 5% rise in death rates in human societies, especially in infections caused by the flu virus [6]. The study of rinderpest-infected rabbits showed that there was an increase in mortality rate, especially when the fever is inhibited using antipyretic drugs such as aspirin (acetylsalicylic acid); 70% of the animals that were treated with this drug died due to infections, unlike the 16% that had a normal body temperature. There is no beneficial outcome of fever in extreme inflammatory responses where the reduction in body temperature serves as a protective mechanism [9]. During febrile conditions, the maintenance of fever is done via a coordinated approach that involves an interplay among the innate immunities as well as the neuronal circuits of the central and peripheral nervous systems. The immune system recognizes the infection within the pathogen–associated molecular patterns of toll-like receptors. Prostaglandin E2 [10], also known as dinoprostone, is produced by lipopolysaccharide (LPS) [11], which stimulates the macrophages to release cytokines, interleukin-1β (IL-1) [11], as well as tumor necrosis factor (TNF) [12]. LPS also activates toll-like receptor 4, a transmembrane protein belonging to the family of pattern recognition receptors. LPS travels through the blood-brain barrier, and then fever is initiated. On the other hand, endotherms elevate body temperatures via the release of noradrenaline, which increases thermogenesis as well as constriction of the extremities [13,14].

Heat shock proteins (HSPs) are continuously produced, and they respond rapidly to proteotoxic stressful conditions, such as heat, hypoxia, inflammatory processes, exposure to toxins, food restriction, and infections. Post-translational modifications (SUMOylation and phosphorylation of heat shock factor protein 1) cause HSPs to be turned on by stress (HSF1) [15]. These changes dissociate HSF1 from its complexity with

Transcription factors and result in the formation of homotrimers that transduce to the nuclei. Despite the direct delivery of interleukins, interleukin (IL-6) or TNF into the brain, it induces a febrile reaction. However, multiple lines of evidence indicate that this is not always the case [16]. During the LPS-induced inflammatory response in mice, passively elevating the core temperature to a higher range utilizing whole-body abnormal temperature (hyperthermia) significantly increases plasma concentrations of IL-1, IL-6, and TNF. Tocilizumab, an IL-six blocker, was discovered to prevent hyperthermia that arises after thymus cell-based immunotherapeutic application in children with juvenile malignancies [17]. Fever also regulates protective immunity during infections. The first to respond to this infective organism are the innate cells, which invade the site of infection hours later, thereby phagocytically destroying the pathogen. The macrophages close the vacuum existing in both adaptive as well as innate immunities. The pathogen is taken up in the peripheral tissues and then relocates to the lymph nodes [18]. The range of temperatures in fever stimulates all the steps, which promotes innate as well as adaptive immunities. In vivo and in vitro studies have shown that fever has primarily been noted by using hyperthymic temperatures in the febrile range for mammals. During febrile conditions, heat conservation differs from healing mechanisms that are created by thermoregulators [19].

18.1.1 Febrile treatment

Most of the time treating fever is not always necessary. The clinical advice is to give the patients analgesics such as paracetamol or ibuprofen, especially when they are not comfortable, irritable, or in pain. Fever helps the body to fight infections, as such reducing the fever may prolongs the illness. If the infection is of viral origin, treatment with antibiotics does not have any positive impact, but antibiotics do a better job in cases of bacterial infections [8]. Meanwhile, there are high temperatures of unknown origin. At this point, medications cannot reduce such conditions. The hypothalamus is primarily responsible for temperature reduction in the body. The rise in temperature serves as a defense mechanism against bacteria or viral pathogens. It could be due to radiological, nutritional, or even psychological effects that result in the temperature rising. When the temperature rises beyond $38°C$, it is called pyrexia of unknown origin [20] In the management of fever, breathing, airways, as well as air circulation are necessary. The patients are given intubation to help breathe in critical cases by observing the cardiac rhythms as well as hypoxia. Intravenous fluids, such as dextrose saline, could be given, temperature measurements should be taken; and the patient's clothing is also removed to create an ideal environment for adequate dissolved oxygen and systematic air flow [21−23]. The temperature is reduced at the rate of 0.1 degrees per minute to reach 39°C. This is achieved through evaporative cooling and cooling with cold water. In evaporative cooling, the patient is sprinkled with warm water while allowing a fan to move air around the entire body. Mechanical evaporative cooling could also be achieved using the air conditioning system. Basically there are three main ways to cool with evaporation: fan-and-pad devices, unit cooling systems (also called swamp coolers), and a moisture control system. In the fan-and-pad method, pumps move water over a permeable or cellulose pad at one end of the greenhouse. The vent blowers at the opposite end suck air from either the exterior through freshly washed pads. A disadvantage of fan-and-pad systems is their expensive maintenance requirements and inconsistency. Evaporative air coolers, often referred to as swamp coolers, are metal-cased, packaged devices that are positioned outside. The padding is composed of organic matter and is continuously saturated by something like a water pump with a recirculation system. A fan sucks in air from the exterior, which then travels through into the pads on three sides and exits through a duct on the fourth [24−26]. The treatment with a cold-water bath overall has a rapid temperature reduction. Usually, the patient is placed in a basin of cold water to take advantage of the rapid conduction of water to heat [24,27,28].

18.1.2 Temperature measurement

Temperature is an extremely important but commonly monitored parameter for the majority of mechanical-engineered devices. Temperatures must be maintained or controlled for numerous procedures. Many processes in life require temperature measurement, and the thermometer is the device for temperature measurement. There are a variety of thermometers in use today, depending on what one wants to measure. Thermometers started as early as 150AD, during the time of Galen, when he observed the complexion of an individual. The science behind thermometry boomed in 1500 [29]. The first thermometer was an air thermoscope; afterward 18 different modifications came into existence [30]. Gabriel Fahrenheit learned how to calibrate thermometers from Danish astronomer Ole Romer [31,32]. Fahrenheit produced thermometers using Romer's scale, and it was also modified to what is today known as the Fahrenheit scale in the years 1708–1724 [33]. His methodology was his trade secret, though some scholars stated that he used the melting of sea salt and armpit water as his calibration point [30]. The UK scientist adopted the temperature, and the 212 scale was recorded as the boiling point of water. This same number, 212, was also used as the melting point of ice [33,34]. In the year 1740, Anders Celsius proposed the centigrade scaling system. The inventor of the scale was not scientifically documented, notwithstanding Anders's work, which defined the melting point of ice as 100 and the steam point as 0. Linnaeus inverted this idea, using 0 as the melting point and 100 as the steam point. The centigrade scaling system was changed to Celsius in the year 1984 [35]. Different varieties of thermometers are on the market today. Most of them are highly precise owing to their fluid properties, especially mercury. The only shortcoming among these is the accuracy and resolution. Many manufacturers have made thermometers with variable scales, such as wet viscosity and glass thermometers. The glass thermometer is usually preferred because of its repeatability. In modern times, apart from mercury thermometer fluids are used. This is because of the hazards associated with mercury spills.

The range and accuracy of thermometers depend on the size of the hole, the tube length, and the fluid in the thermometer. The reading side with a rounded shape serves as a magnifying glass, thus making the column readable [36]. The accuracy of the thermometer also depends on the process that is affected by daily usage. In industries bimetal thermometers are used. The accuracy of this thermometer is less, though more rugged compared with glass thermometers. They are mainly used in industrial settings and are constructed using a metallic rod with a sensor. Another type of thermometer is the bimetal thermostats [37]. These thermometers are affected by changes in low or higher temperature values. The two-metal thermometer (bimetal and Bourdon tube) is the most common and consists of two different metallic strips, such as iron and copper, which are fastened together to form one rod. Both metals expand with an increase in temperature, but the rate of expansion of each differs from the other, which makes the rod bend. The curvature of the rod causes a deflection in the pointer, thereby indicating a temperature change [30]. One of the most popular types of thermometers is the bimetal thermometer, which contains a pen to record temperature changes. Again, electrical thermometers are another type of thermometer. Resistance and thermocouples are two examples of electrically made thermometers. Thermocouples have two different metallic wires whose ends are twisted together to form two connections — the reference link at a constant temperature (0°C) [36]. The metal wires (Constantan) of most thermocouples are used in measuring the air temperature. As for the thermocouple wires used in measuring high temperatures, which may reach 2800°C, they are made of metals that can withstand high temperatures [38]. In these thermometers, the change in the electrical resistance of the metal arising from the temperature change is measured, and the change in resistance is translated into degrees of heat. The electrical thermometer can as well be in digit form, whose reading of the temperature is shown in the form of numbers. Digital thermometers measure temperatures through a precise instrument called a probe. The probe is made either from a metallic material such as copper or platinum or from one of the semiconducting materials. Temperature changes cause large changes in the electrical resistance of these materials. Meanwhile, some thermometers can be used singly. They are less expensive to manufacture than ordinary thermometers.

18.2 Methodology

The methodology involves the device and method employed to assemble the parts so that they function properly (Figs. 18.1 and 18.2).

18.2.1 Control circuit

This is usually an Arduino whose function is to provide a closing or opening circuit. The Arduino is an open-source electronic gadget used in the development of ideas and projects. The board recognizes the programming language Arduino IDE (Arduino Integrated Development Environment), which can be downloaded for free from the web. The open hardware allows for modifications in the engineering designs. The codes can be reprogrammed completely to suit individual needs. The Arduino C programming language features are completely free and not similar to the programming language of the environmentalists, such as Micro C, which requires a huge amount of money to purchase. The control unit is very different from other microcontroller development boards because of the simplicity of the programming language. The Arduino

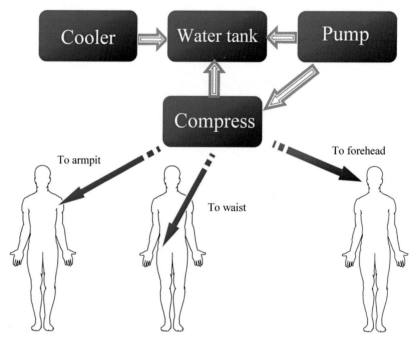

Figure 18.1 Body parts to apply temperature reducer machine.

programming language was derived from the processing language, which is the basis of modern programming languages and the revolution in software technology. Arduino is not designed only for amateurs, it has been developed to suit all levels, ranging from amateurs to advanced projects, and the evidence is that there are great features that make Arduino at the top of microcontrollers, including the possibility of integrating into projects that are programmed in advanced engineering languages such as Matlab and Java, where you will find ready-made software libraries, especially for dealing with Arduino

In the project design, a tank that contains water is used in the cooling process. This tank is made of steel sheets with a thickness of 1 mm and dimensions of $10 \times 10 \times 15$ cm. This tank was constructed by the welders using electric arc welding, the metal of the tank is stainless steel. The welding aims to obtain a nondisintegrating connection, and this is done either by local or complete melting of the edges or by causing plastic agitations in these two surfaces without heating. In recent times, welding is considered an atmosphere of protective gases for welding structures made of different types of carbon and alloy steels and constructions made of nonferrous metals and

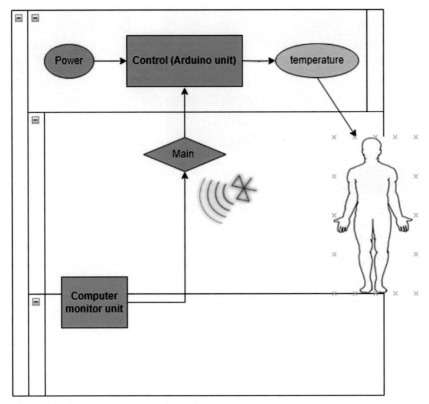

Figure 18.2 External structure of the control unit.

their alloys. This welding process can be done manually, semiautomatically, or automatically. The welders use a protective inert gas (argon and helium). The most commonly used gases at present are argon and carbon dioxide gas, or a mixture of them. This type of welding method is distinguished from other methods of electric arc welding in that the gas protects the arc and the molten metal pool from the harmful influence of the surrounding medium (air, for example). The tank was insulated using ArmaFlex insulation, which is a foam material composed of closed cells that do not contain fibers. It is equipped with a smooth layer either on one side or on both sides, and it forms the outer surface of the insulation, while the inner closed cells form the effective insulation material. These materials are manufactured without Chlorofluorocarbons or hydrochlorofluorocarbons. It contains antimicrobial agents that prevent rotting and reduce noise. The thermoelectric cooler is then installed (Fig. 18.3).

Figure 18.3 Insulation of the tank.

The thermoelectric cooler used in this project is TECI 12706. These coolers are a form of solid-state refrigeration, including semiconductor and electronics assembly. The cooler is solid and lacks vibration or noise, so it is easy to install and operate because it is mainly semiconductors located between the ceramic plates and has no moving parts.

18.2.1.1 Heat exchanger

A heating element is a part that moves liquids through all the ducts that are permeated by another channel. This changes the temperature of the liquid. If we want to boost the temperature of a fluid, we need to use an exceptionally hot medium. A fluid that needs to be refrigerated can also be cooled by running through pipes that pass through a reduced medium. Heat exchange refers to the method of moving heat from one substance to another. A heating element is the thing that makes this process happen. For instance, an air conditioning unit is a heating element that lowers the temperature within a room by sending air through the pipelines that convey cold refrigerant vapor (Freon).

18.2.1.2 Temperature sensor (LM35)

It is a small integrated circuit that gives its output an electrical voltage proportional to its degree of temperature. This sensor is available in the form of a transistor with the packaging of TO-92. When the sensor is connected to the direct method, the sensor measures the temperature values from $+2°C$ to $150°C$. The sensor's supply voltage must be between 4 V and 30 V, so it can be fed directly from the 5 V used in the Arduino circuit. The sensor output gives a constant voltage signal whose value is proportional to the measured temperature as follows: 10 mV or 0.01 V/°C.

18.3 Conclusion

The device is made with iron, which cools the water and acts as a reservoir for the cooled water to be pumped. The device is designed with an Arduino control system, an insulator, and a thermoelectric insulator. The heat charger changes the temperature of fluids, and the temperature sensor responds to the variation in temperature. The device converts the thermal signal to the setpoint, which is the required temperature, and afterward turns on the last control mechanism. The last control system changes the parameter that is being changed to adjust the quantity of heat being made or taken from the procedure.

References

[1] A.K. Mishra, M.G.L.C. Loomans, J.L.M. Hensen, Thermal comfort of heterogeneous and dynamic indoor conditions—an overview, Building and Environment 109 (2016) 82–100. Available from: https://doi.org/10.1016/J.BUILDENV.2016.09.016.
[2] N. Body, Unit: XIII: Body temperature regulation, and fever, Metabolism and Temperature Regulation, Chapter 73, 1948.
[3] H. Yousef, E.R. Ahangar, M. Varacallo, Physiology, thermal regulation, StatPearls (2022)Accessed: Jul. 04, 2022. [Online]. Available. Available from: https://www.ncbi.nlm.nih.gov/books/NBK499843/.
[4] Human body temperature has decreased in the United States, study finds | News Center | Stanford Medicine. https://med.stanford.edu/news/all-news/2020/01/human-body-temperature-has-decreased-in-united-states.html (accessed July 04, 2022).
[5] M. Protsiv, C. Ley, J. Lankester, T. Hastie, J. Parsonnet, Decreasing human body temperature in the United States since the industrial revolution, Elife 9 (2020). Available from: https://doi.org/10.7554/ELIFE.49555.
[6] S.S. Evans, E.A. Repasky, D.T. Fisher, Fever and the thermal regulation of immunity: the immune system feels the heat, Nature Publishing Group (2015). Available from: https://doi.org/10.1038/nri3843.

[7] J.J.G. Plaza, N. Hulak, Z. Zhumadilov, A. Akilzhanova, Fever as an important resource for infectious diseases research, Intractable & Rare Diseases Research 5 (2) (2016) 97. Available from: https://doi.org/10.5582/IRDR.2016.01009.

[8] L. Geddes, The fever paradox, New Scientist (1971) 246 (3277) (2020) 39. Available from: https://doi.org/10.1016/S0262-4079(20)30731-4.

[9] A.M.P. Schieber, J.S. Ayres, Thermoregulation as a disease tolerance defense strategy, Pathogens and Disease 74 (9) (2016) 106. Available from: https://doi.org/10.1093/FEMSPD/FTW106.

[10] K.H. Parker, D.W. Beury, S. Ostrand-Rosenberg, Myeloid-derived suppressor cells: critical cells driving immune suppression in the tumor microenvironment, Advances in Cancer Research 128 (2015) 95−139. Available from: https://doi.org/10.1016/BS.ACR.2015.04.002.

[11] H.Y. Hsu, M.H. Wen, Lipopolysaccharide-mediated reactive oxygen species and signal transduction in the regulation of interleukin-1 gene expression, Journal of Biological Chemistry 277 (25) (2002) 22131−22139. Available from: https://doi.org/10.1074/JBC.M111883200.

[12] W. Tong, et al., Resveratrol inhibits LPS-induced inflammation through suppressing the signaling cascades of TLR4-NF-κB/MAPKs/IRF3, Experimental and Therapeutic Medicine 19 (3) (2020) 1824−1834. Available from: https://doi.org/10.3892/ETM.2019.8396.

[13] J.K. Hermann, et al., The role of toll-like receptor 2 and 4 innate immunity pathways in intracortical microelectrode-induced neuroinflammation, Frontiers in Bioengineering and Biotechnology 6 (2018) 113. Available from: https://doi.org/10.3389/FBIOE.2018.00113/BIBTEX.

[14] A. Tohidpour, et al., Neuroinflammation and infection: molecular mechanisms associated with dysfunction of the neurovascular unit, Frontiers in Cellular and Infection Microbiology 7 (2017) 276. Available from: https://doi.org/10.3389/FCIMB.2017.00276/BIBTEX.

[15] K.H. Su, C. Dai, Metabolic Control of the proteotoxic stress response: implications in diabetes mellitus and neurodegenerative disorders, Cellular and Molecular Life Sciences. CMLS 73 (22) (2016) 4231. Available from: https://doi.org/10.1007/S00018-016-2291-1.

[16] Heat Shock Factor. https://www.sdbonline.org/sites/fly/dbzhnsky/heatsf1.htm (accessed July 04, 2022).

[17] D. Mishra, et al., Parabrachial interleukin-6 reduces body weight and food intake and increases thermogenesis to regulate energy metabolism, Cell Reports 26 (11) (2019) 3011−3026.e5. Available from: https://doi.org/10.1016/J.CELREP.2019.02.044.

[18] S.D. Rosenzweig, S.M. Holland, Recent insights into the pathobiology of innate immune deficiencies, Current Allergy and Asthma Reports 11 (5) (2011) 369−377. Available from: https://doi.org/10.1007/S11882-011-0212-9.

[19] K. Rakus, M. Ronsmans, A. Vanderplasschen, Behavioral fever in ectothermic vertebrates, Developmental and Comparative Immunology 66 (2017) 84−91. Available from: https://doi.org/10.1016/j.dci.2016.06.027.

[20] A.D. Inchingolo, et al., Benefits and implications of resveratrol supplementation on microbiota modulations: a systematic review of the literature, International Journal of Molecular Sciences 23 (7) (2022). Available from: https://doi.org/10.3390/ijms23074027. MDPI.

[21] A. Garba Usman, M. Alhosen, A. Alsharksi Çankırı Karatekin Üniversitesi, A. Muhammed Naibi, Applications of Artificial Intelligence-Based Models and Multi-Linear Regression for the Prediction of Thyroid Stimulating Hormone Level in the Human Body. Available: https://www.researchgate.net/publication/342571024.

[22] S.I. Malami, et al., Implementation of soft-computing models for prediction of flexural strength of pervious concrete hybridized with rice husk ash and calcium carbide

waste, Modeling Earth Systems and Environment 8 (2) (2021) 1933−1947. Available from: https://doi.org/10.1007/S40808-021-01195-4.

[23] S.I. Abba, et al., Comparative implementation between neuro-emotional genetic algorithm and novel ensemble computing techniques for modeling dissolved oxygen concentration, *Hydrological Sciences Journal*, 66:10, 1584-1596, DOI: 10.1080/02626667.2021.1937179.

[24] I. Ozsahin, M.T. Mustapha, S. Albarwary, B. Sanlidag, D.U. Ozsahin, T.A. Butler, An investigation to choose the proper therapy technique in the management of autism spectrum disorder, Journal of Comparative Effectiveness Research 10 (5) (2021) 423−437. Available from: https://doi.org/10.2217/CER-2020-0162.

[25] M.T. Mustapha, D.U. Ozsahin, I. Ozsahin, B. Uzun, Breast cancer screening based on supervised learning and multi-criteria decision-making, Diagnostics 12 (6) (2022) 1326. Available from: https://doi.org/10.3390/diagnostics12061326.

[26] H. Gokcekus, D.U. Ozsahin, M.T. Mustapha, Simulation and evaluation of water sterilization devices, Desalination and Water Treatment 177 (2020) 431−436. Available from: https://doi.org/10.5004/dwt.2020.25115.

[27] Canada. Health Canada., A. and C. C. Bureau. Canada. Water, and Canadian Electronic Library, Extreme heat events guidelines, a technical guide for health care workers, 2012, p. 149.

[28] Participant Workbook.

[29] Temperature Measurement.

[30] The history of the thermometer. https://www.thoughtco.com/the-history-of-the-thermometer-1992525 (accessed July 05, 2022).

[31] The thermometer-from the feeling to the instrument. https://www.scribd.com/document/459012090/The-Thermometer-From-The-Feeling-To-The-Instrument (accessed July 05, 2022).

[32] The strange history of the invention of the thermometer, Time. https://time.com/6053214/thermometer-history/ (accessed July 05, 2022).

[33] E. Grodzinsky, M.S. Levander, History of the thermometer, Understanding Fever and Body Temperature (2020) 23. Available from: https://doi.org/10.1007/978-3-030-21886-7_3.

[34] Temperature scales. https://www.montereyinstitute.org/courses/DevelopmentalMath/COURSE_TEXT_RESOURCE/U06_L3_T1_text_final.html (accessed July 05, 2022).

[35] F. Domínguez-Castro, J.M. Vaquero, M. Marín, M.C. Gallego, R. García-Herrera, How useful could Arabic documentary sources be for reconstructing past climate? Weather 67 (3) (2012) 76−82. Available from: https://doi.org/10.1002/WEA.835.

[36] D. Camuffo, Measuring temperature, Microclimate for Cultural Heritage (2014) 395−432. Available from: https://doi.org/10.1016/B978-0-444-63296-8.00012-3.

[37] Measuring Temperature Accurately: What Are the Costs? from Cole-Parmer United Kingdom. https://www.coleparmer.co.uk/tech-article/measuring-temperature-accurately-what-are-the-costs (accessed July 05, 2022).

[38] Temperature sensor, Hi Temp Solutions. http://hitempsolutions.com/temperature-sensor.aspx (accessed July 05, 2022).

CHAPTER NINETEEN

Development of a brain–computer interface device converting brain signals to audio and written words

Dilber Uzun Ozsahin[1,2,3], Basil Bartholomew Duwa[3,4], Abdelrahman Himaid[4], Declan Ikechukwu Emegano[3,4] and Ilker Ozsahin[3,5]

[1]Department of Medical Diagnostic Imaging, College of Health Science, University of Sharjah, Sharjah, United Arab Emirates
[2]Research Institute for Medical and Health Sciences, University of Sharjah, Sharjah, United Arab Emirates
[3]Operational Research Center in Healthcare, Near East University, Nicosia/TRNC, Mersin 10, Turkey
[4]Department of Biomedical Engineering, Near East University, Nicosia/TRNC, Mersin 10, Turkey
[5]Department of Radiology, Brain Health Imaging Institute, Weill Cornell Medicine, New York, NY, United States

Contents

19.1 Introduction

According to the National Institute on Deafness and Other Communication Disorders, individuals with speech impediments make up between 6% and 8% of the overall population of the United States (328.2 million people). According to a global population estimate, over 170 million

Practical Design and Applications of Medical Devices.
DOI: https://doi.org/10.1016/B978-0-443-14133-1.00021-5

people worldwide have some form of speech disability [1]. Assistive equipment and interpreting services may be needed by millions of people who are unable to communicate successfully on their own. Similarly, according to a survey published by the World Health Organization, by the year 2050, it is anticipated that approximately 2.5 billion people will have some degree of hearing loss, and approximately 700 million will need hearing rehabilitation in some cases [2].

Currently, devices and other improved speech assistive gadgets are being developed to assist speech-impaired individuals, which employ intelligent models. Similarly, the National Institute of Health conducted a study to investigate recent developments in brain technologies using artificial intelligence. In their research, they were able to accurately replicate voice sounds using brain impulses that were recorded from people who had epilepsy. As a result, the patients were given reading assignments consisting of complete phrases, and the associated data from their brains were scanned to power computer-generated speeches. In addition, the computer was able to generate a variety of identical sounds simply by having the user mimic the act of speaking. This knowledge was sufficient to the computer [3].

Patients who have experienced face, tongue, or laryngeal muscle paralysis as a result of a stroke or other related neurological illnesses may find that the inability to communicate has a debilitating effect on their quality of life. Devices that interpret movement or visual acuity into voice have been developed thanks to artificial intelligence technology, which has assisted these individuals in communicating. As a result of the fact that these systems require the selection of certain letters or even full phrases to construct sentences, the maximum possible processing speed for them is extremely restricted [4].

A similar study by Nicole Martin, published in the journal *AI and big data* revealed a study that could help folks who have lost their voices. A system that can convert brain signals into speech has been developed by these scientists using electrodes and advanced artificial intelligence. People who have suffered brain trauma or neurological conditions such as epilepsy, Alzheimer's disease, multiple sclerosis, Parkinson's disease, and others may benefit from this technique [5].

Using brain signals as input, the machine acts and selects words that it thinks are being analyzed in the brain at the time. To better understand the English language, the researchers utilized a computer to transcribe spoken utterances and select the best possible term from a list of possible ones. As a next step, they employed brain impulses that could be traced to the human mouth, jaw, tongue, and throat. These were then utilized to anticipate words by the machine learning algorithms. Previously, they were employed to aid those who were paralyzed from using their hands to type [6].

In recent years, one of the most imperative areas that has been the subject of research and study is the brain–computer interface (BCI). This branch of study originates from the brain waves that are released by the brain, as it depends on these signals to establish a line of communication between the neural network and the computer. There is no question that the attention that has been given in this sector is a direct result of the excellent efforts and remarkable achievements that have been made to revitalize life in all of its many facets, most notably healthy living. It is enough for you to read a number of the outcomes of this field to grasp the enormous need for its continued growth and to become aware of its relevance [7].

A BCI is a computer-based device that can receive brain signals, evaluate those signals, and convert those analyses into commands that may be transmitted to an output device to carry out a particular action that the user desires. Therefore, the typical output channels of the brain, which include the peripheral nerves and muscles, are not utilized by BCIs. According to this definition, the phrase BCI can only refer to devices that monitor and process the signals generated by the central nervous system (CNS) [8]. Therefore, the typical output channels of the brain, which include the peripheral nerves and muscles, are not utilized by BCIs. According to this definition, the phrase BCI can only refer to devices that monitor and process the signals generated by the CNS. Therefore a communication system that is activated by the user's voice or their muscles is not an example of a BCI. In addition, an electroencephalogram (EEG) equipment by itself is not a BCI because it just records brain signals and does not produce an output that operates on the surroundings of the user. In general, BCI systems can be categorized in a number of different ways. The three different classification systems are mode of operation, dependability, and imaging. In terms of their level of dependability, BCIs can be divided into two categories: dependent and independent [9]. Dependent BCIs are the ones that let people employ some type of motor control, such as gaze, to communicate with the device. Examples of dependent BCIs that are often used include ones that are based on motor imagery. Independent BCIs, on the other hand, do not require any motor control and are therefore ideally suited for individuals who have suffered a stroke or who suffer from locked-in syndrome.

BCIs can be invasive or noninvasive, depending on the imaging techniques used. Noninvasive techniques include magnetoencephalography, electroencephalography, near-infrared spectroscopy, positron emission tomography, and functional magnetic resonance imaging, whereas invasive techniques include electrocardiograms and single-neuron recordings. The BCI can function in either a synchronous or asynchronous mode depending on the situation. A synchronous BCI requires the subject to communicate with the system in response to a cue that is imposed by the system over the course of a predetermined time period. A subject using an asynchronous BCI, on the other hand, is granted the ability to interface with the program at any point in time. The usability of synchronous BCIs is lower than that of asynchronous BCIs; nonetheless, synchronous BCIs are simpler and easier to construct [10].

19.2 Related studies

The history of BCI dates back to the period when Hans Berger was credited with discovering the electrical activity of the human brain as well as inventing electroencephalography, both of which are important milestones in the development of BCIs. Both of these occurrences are taken into consideration as the starting points of EEG. Berger is credited with being the pioneer scientist to study the activity of the human brain using EEG in the year 1924. Berger was able to identify rhythmic activity in the brain by examining EEG traces, and he labeled several of these patterns after him, namely the alpha and Berger's wave. Berger's discovery allowed for the discovery of rhythmic activity in the brain (8−13 Hz) [11].

Research on BCIs was initiated at the University of California in the 1970s, which led to the coining of the term "brain−computer interface." BCI research and development are still predominantly focused on neuroprosthetics applications, which can help restore impaired sight, hearing, and movement. This is expected to be the case for the foreseeable future. In the middle of the 1990s, scientists successfully implanted humans with their very first neuroprosthetic devices. The BCI does not correctly read the mind but rather identifies the tiniest of shifts in the energy radiated by the brain when the user thinks in a particular way. A BCI is able to identify particular energy and frequency patterns in the brain [12].

19.2.1 Classification of brain−computer interface

The following are the three primary categories that can be used to classify BCIs:

Fig. 19.1 shows the schematic representation of the categorized forms of BCI. In invasive BCI, specialized equipment is utilized to record data (brain signals), and this equipment must first be surgically implanted into a human subject in the course of the operation. Contrarily, in semiinvasive BCI, devices are placed at the top of the human brain's skull. Devices that are noninvasive are typically thought to be the safest and least expensive option available. However, as a result of the barrier caused by the skull, these devices are only capable of capturing weaker human brain impulses. Electrodes positioned on the scalp allow for the detection of electrical activity emanating from the brain [13].

19.2.1.1 BCI system process

In humans, the beginning of the electrical activity of the brain typically occurs between three and seven weeks before birth. This activity can be defined as an electrical process that is collected either directly from the surface of the brain or indirectly from the surface of the brain's exterior. The brain is responsible for the vibration, and the strength of this electrical activity as well as its form are both determined by whether or not the electrical current is continuously present. Mostly on the stimulation and excitation that come about as a direct result of the various functions that are carried out by the stimulation system. The retina of the brain stem is where brain waves are generated, and the ripples in this electrical potential are what gave rise to the term "brain waves" [14].

EEG signals in newborns have a greater frequency, δ and θ waves, and a range of (30−60) microvolts. However, in adults, these waves are of higher frequency, and their amplitude increases with the passage of time; this is

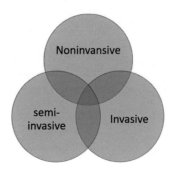

Figure 19.1 Classification of brain−computer interface.

referred to as the brain maturation phase. Although the existence of high-frequency alpha waves is associated with a more stable electrical image of the brain as we age, this phenomenon is only observed in young people.

The amplitude and frequency of the brain's right and left electrical signals must be in harmony. The presence of brain disease is indicated if there is a discrepancy between these symptoms of more than 30%. Changes in brain signals occur over time or with age, and depending on the patient's condition (concentration state, mentally distracted state, forgetfulness state, sleep state, need for oxygen, degree of hunger, degree of thirst, degree of saturation), due to biochemical and other causes, whereas hormonal and neurological causes [15].

19.3 Method

19.3.1 Experimental design (Fig. 19.2)

19.3.2 Components

19.3.2.1 EEG preprocessing

Electrical potentials are produced as a result of the neurological function of brain cells, and it is possible to examine these electrical potentials. Amplification and citation methods that are suitable for the situation are

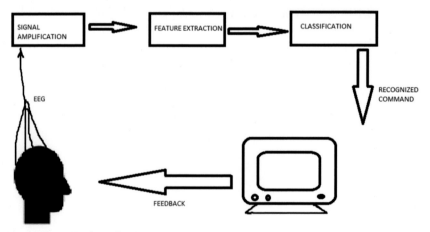

Figure 19.2 EEG-based BCI structure.

utilized to monitor and quote it directly from the scalp. The EEG signal is what is being referred to here, however, this signal is what reflects the functions. It is not a particular cognitive activity but rather an aggregate of the brain that uses the quoting and processing system. During a session that lasts 10—20 minutes, neurologists use this method to capture the electrical activity that occurs in a patient's brain. The patient closes his eyes and settles into a comfortable position on a chair provided by the medical facility [16].

19.3.2.1.1 Signal amplification

The act of measuring the analog signals originating from the brain by employing a variety of different sensors is referred to as signal acquisition. To get rid of the noise in the signal, the received signal is amplified and filtered after it has been received. In the final step, an analog-to–digital converter is used to transform the signal into a digital format before it is sent to the processing unit [17].

19.3.2.1.2 Feature extraction

The process of obtaining distinctive characteristics from a received signal is referred to as feature extraction. These features ought to have strong correlations with the user's intent to function properly. Because a significant portion of the relevant brain activity—that is, the activity that is most strongly correlated—is either transitory or oscillatory [18].

19.3.2.2 Feature translation

The classification module, which is often referred to as the translation algorithm, is the most important component of every BCI. The user's electrophysiological input is simply converted into an output that controls the devices that are external to the body. The translation algorithm is an essential part of the signal-processing module of the BCI system. Its job is to convert the signal properties that were extracted into device commands that will carry out the user's intended action. No matter what form it takes, a translation algorithm will transform signal characteristics into device control commands. The very first stage of signal processing consists of nothing more than the extraction of certain signal characteristics. Both the frequency and shape characteristics of the extracted signal can be categorized using either linear or nonlinear approaches, such as neural networks, depending on the type of analysis being performed [19].

19.3.2.3 Device output

The output device receives the output of the feature translation unit once it has been processed. The commands that come from the feature translation algorithm are what operate the external device. These commands provide functionality to the device, such as the ability to choose letters. The user is provided with feedback through the operation of the device, which completes the control loop [20].

19.3.2.4 Signal processing

The stage of signal analysis and feature extraction is one of the most significant phases since it is where the signal features will be used to determine the identity of the signal and where the signals will be differentiated based on those features. The features will be stored in an array that is going to be referred to as the feature matrix, and each signal that is handled will have its own matrix [21].

19.3.2.5 Crest, trough, and amplitude

The highest point of a wave is referred to as the crest, while the lowest point is referred to as the trough. The amplitude can be determined by first determining the difference in value that exists between the highest and lowest points. Before continuing with these computations, it is necessary to first identify the minimum and maximum values of the signal. As can be seen in Fig. 19.3, both the upper and lower boundaries exhibit a divergence of 2% for the upward edge [22].

There is a connection between each example and the lower and upper boundaries. Given the amount of tolerance in the lower and higher limits with small values, such as 2/100 or 3/100, in general, these limits are established by the difference between the highest and lowest values. The area of tolerance is defined, and the relationship between the two variables gives the amount of variance for the lower bounds.

19.4 Results

19.4.1 Working algorithm

To estimate the values in binary, the histogram approach is utilized, and the process can be summed up as follows:

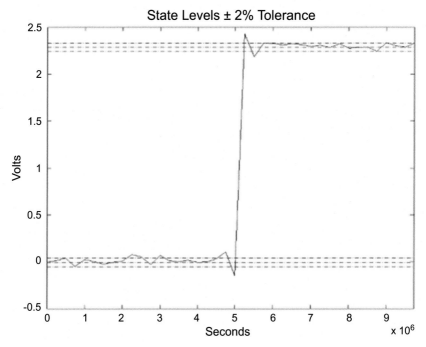

Figure 19.3 The upper and lower limits.

- The box size is defined as the proportion of the amplitude field to the number of histogram boxes, and this ratio is applicable to a particular number of histogram boxes.
- Placing the values of the data into the appropriate boxes on the histogram.
- During calculating the nonzero values, determine the box with the lowest value, denoted by ilow, and the box with the greatest value, denoted by ihigh.
- Split the histogram into two different subsections, as follows:
  - The histogram with the lowest values is 1-ilowi1/2(ihigh-ilow).
  - The histogram with the value 2-.ilow + 1/2(ihigh-ilow)iihigh is the highest possible value.

19.4.2 Rise time on average and edges with positive polarity (Rising)

The amount of time it takes, on average, to go from the level of the lowest reference to the level of the highest reference is what is referred to as the average rise time. The positive transformation of the first derivative, expressed as dx/dt greater than zero, denotes the upward edge of the graph. The beneficial development is depicted in Fig. 19.4. It has come to our attention that the amplitude of the wave is not a factor in the process of detecting the rising pulse; rather, what matters is the direction in which the wave is changing.

19.4.3 Time spent landing on average and edges with negative polarity (Falling)

The amount of time it takes, on average, to descend from the highest reference level to the lowest reference level is referred to as the average landing time. The descending edge is denoted by a negative transformation of the first derivative, which is written as dx/dt less than zero. The detrimental shift is depicted in Fig. 19.5. It is clear that the detection of the pulse is not connected to the amplitude of the change but rather to the direction in which it is occurring.

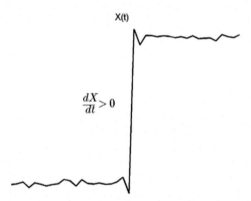

Figure 19.4 Positive shift of the first derivative of the impulse.

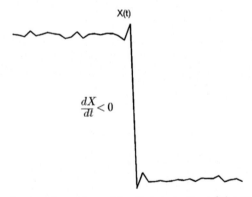

Figure 19.5 Negative transformation of the first derivative of the impulse.

19.4.3.1 Simulation

```
%      N_SL=0;
%else
      N_SL=mean(S_ne)/1000;
%end

% 2 - Cycles:
%Period:
P = pulseperiod(s,fs);
P_ms=mean(P)*1000;
%Frequency:
F_HZ=1/mean(P);

%+Pulses:
W = pulsewidth(s,fs);
P_P=length(W);
%-Pulses:
N_P=length(W);

%+Width:
P_Wi=mean(W)*1000;
%-Width:
N_Wi=(P_ms-P_Wi);

%+Duty Cycle:
D = dutycycle(s,fs);
P_DC=mean(D)*100;
%-Duty Cycle:
N_DC=100-P_DC;

%Feature Matrics:
Am = single (Am);
%['Amplitude';'+Edge';'+RiseTime';'+SlewRate';'-Edge';'-
FallTime';'-Slew
Rate';'Period';'Frequancy';'+Pulses';'+Width';'+DutyCycle
';'-Pulses';'-Width';'-DutyCycle'];
y=[Am;P_E;P_RT;P_SL;N_E;N_FT;N_SL;P_ms;F_HZ;P_P;P_Wi;P_DC
;N_P;N_Wi;N_DC];

x1='Amplitude';x2='P_Edge';x3='RiseTime';x4='P_SlewRate';
x5='N_Edge';x6='FallTime';x7='N_SlewRate';x8='Period';x9=
'Frequancy';
x10='P_Pulses';x11='P_Width';x12='P_DutyCycle';x13='N_Pul
ese';x14='N_Width';x15='N_DutyCycle';
%Structrue:
X=struct(x1,y(1),x2,y(2),x3,y(3),x4,y(4),x5,y(5),x6,y(6),
x7,y(7),x8,y(8),x9,y(9),x10,y(10),x11,y(11),x12,y(12),x13
,y(13),x14,y(14),x15,y(15));
```

```
% create a neural network
net=feedforwardnet(20);
% train net
net.divideParam.trainRatio = 1; % training set [100%]
net.divideParam.valRatio = 0; % validation set [%]
net.divideParam.testRatio = 0; % test set [%]
% train a neural network

[net,tr,Y,E] = train(net,in,ta);
% show network
%view(net)

% plot network output
figure;
subplot(211)
plot(ta')
title('Targets')
ylim([-2 2])
grid on
subplot(212)
plot(Y')
title('Network response')
xlabel('# sample')
ylim([-2 2])
grid on

% Plots
% Uncomment these lines to enable various plots.
%figure, plotperform(tr)
%figure, plottrainstate(tr)
%figure, plotconfusion(targets,outputs)
%figure, plotroc(targets,outputs)
%figure, ploterrhist(errors)
```

```
m=max(a);
if m==a(1)
    need='water'
    play(s_water)
elseif m==a(2)
    need='hunger'
    play(s_hungry)
elseif m==a(3)
    need='pain'
    play(s_pain)
end
%water_r = audioplayer(water,fs);
%play(water_r);
```

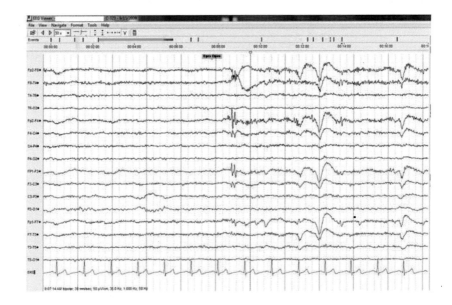

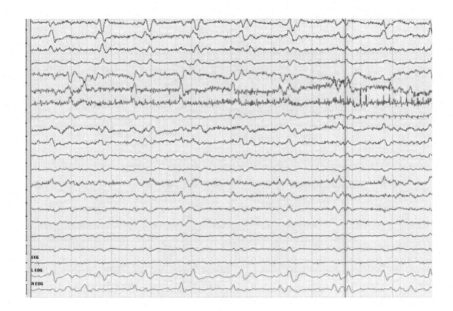

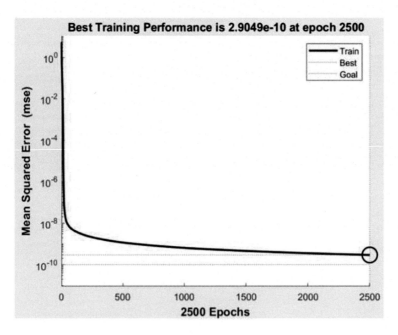

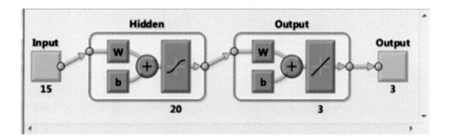

19.4.3.2 Prototype

19.5 Conclusion

The most significant breakthroughs in BCI research and development are still almost entirely relegated to the confines of the laboratory, and the vast majority of the work done to this point consists of data collected from people or animals. Studies on the final target group of people with severe disabilities have primarily been restricted to a small number of limited trials that have been carefully supervised by research professionals. The exciting laboratory development is only just beginning to be translated into clinical use, specifically in the form of BCI systems that can genuinely improve the lives of people with impairments in their day-to-day activities. Therefore this study exposes the knowledge behind the development of a modern BCI that translates into written words on a gadget. However, this device is not a total replacement of the contemporary device, but a miniaturized practical experimentation that simplifies the device.

References

[1] National institute on deafness and other communication disorders, Quick statistics about voice, speech, language, 2022. https://www.nidcd.nih.gov/health/statistics/quick-statistics-voice-speech-language.
[2] World Health Organization, Deafness and hearing loss, 2021. https://www.who.int/news-room/fact-sheets/detail/deafness-and-hearing-loss.
[3] W. Koroshetz, J. Gordon, A. Adams, A. Beckel-Mitchener, J. Churchill, G. Farber, et al., The state of the NIH BRAIN initiative, Journal of Neuroscience 38 (29) (2018) 6427−6438.
[4] A. Gonfalonieri, A beginner's guide to brain-computer interface and convolutional neural networks, 2018.
[5] Nicole Martin, Scientists use ai to turn brain signals into speech, AI & BIG DATA, 2019. https://www.forbes.com/sites/nicolemartin1/2019/05/22/scientists-use-ai-to-turn-brain-signals-into-speech/?sh = 7738241939e5.
[6] G. Karthikeyan, N. Sriraam, Development of brain-computer interface (BCI) model for real-time applications using DSP processors, in: 2009 4th International IEEE/EMBS Conference on Neural Engineering, 2009, pp. 419−422. doi: 10.1109/NER.2009.5109322.
[7] R.S. Mahamune, S.H. Laskar, A review on artefacts removal techniques for electroencephalogram signals, in: 2019 2nd International Conference on Innovations in Electronics, Signal Processing and Communication (IESC), IEEE, 2019, pp. 50−53.
[8] F. Dehrouye-Semnani, N.M. Charkari, S.M.M. Mirbagheri, Toward an improved BCI for damaged CNS-tissue patient using EEG-signal processing approach, arXiv preprint arXiv 2111 (2021) 00757.
[9] S. Vaid, P. Singh, C. Kaur, EEG signal analysis for BCI interface: a review, 2015 Fifth International Conference on Advanced Computing & Communication Technologies, IEEE, 2015, pp. 143−147.
[10] P. Gaur, H. Gupta, A. Chowdhury, K. McCreadie, R.B. Pachori, H. Wang, A sliding window common spatial pattern for enhancing motor imagery classification in EEG-BCI, IEEE Transactions on Instrumentation and Measurement 70 (2021) 1−9.
[11] V. Straebel, W. Thoben, Alvin Lucier's music for solo performer: experimental music beyond sonification, Organised Sound 19 (1) (2014) 17−29.
[12] J.R. Wolpaw, E.W. Wolpaw, Brain-computer interfaces: something new under the sun, Brain-Computer Interfaces: Principles and Practice 14 (2012).
[13] J. Oh, G. Kim, B.G. Nam, H.J. Yoo, A 57 mW 12.5 μJ/Epoch embedded mixed-mode neuro-fuzzy processor for mobile real-time object recognition, IEEE Journal of Solid-State Circuits 48 (11) (2013) 2894−2907.
[14] J.J. Shih, D.J. Krusienski, J.R. Wolpaw, Article highlights, in: Mayo Clinic Proceedings, vol. 3, no. 87, 2012, pp. 268−279.
[15] D.J. McFarland, W.A. Sarnacki, J.R. Wolpaw, Electroencephalographic (EEG) control of three-dimensional movement, Journal of Neural Engineering 7 (3) (2010) 036007.
[16] M.T. Mustapha, D.U. Ozsahin, I. Ozsahin, Comparative evaluation of point-of-care glucometer devices in the management of diabetes mellitus, Applications of Multi-Criteria Decision-Making Theories in Healthcare and Biomedical Engineering, Academic Press, 2021, pp. 117−136.
[17] E.W. Sellers, T.M. Vaughan, J.R. Wolpaw, A brain-computer interface for long-term independent home use, Amyotrophic Lateral Sclerosis 11 (5) (2010) 449−455.

[18] D.J. Krusienski, J.J. Shih, Control of a visual keyboard using an electrocorticographic brain—computer interface, Neurorehabilitation and Neural Repair 25 (4) (2011) 323—331.

[19] J. Mellinger, G. Schalk, C. Braun, H. Preissl, W. Rosenstiel, N. Birbaumer, et al., An MEG-based brain—computer interface (BCI), Neuroimage 36 (3) (2007) 581—593.

[20] S. Yang, R. Li, H. Li, K. Xu, Y. Shi, Q. Wang, et al., Exploring the use of brain-computer interfaces in stroke neurorehabilitation, BioMed Research International 2021 (2021).

[21] X. Gu, Z. Cao, A. Jolfaei, P. Xu, D. Wu, T.P. Jung, et al., EEG-based brain-computer interfaces (BCIs): a survey of recent studies on signal sensing technologies and computational intelligence approaches and their applications, IEEE/ACM Transactions on Computational Biology and Bioinformatics 18 (5) (2021) 1645—1666.

[22] M.H.H.N. Reddy, Brain computer interface drone, Brain-Computer Interface, IntechOpen, 2021.

CHAPTER TWENTY

A dual biventricular resynchronized pacemaker with a remote monitoring system

Dilber Uzun Ozsahin[1,2,3], Basil Bartholomew Duwa[3,4], David Edward[5], Dawda Cham[4], John Bush Idoko[6] and Ilker Ozsahin[3,7]

[1]Department of Medical Diagnostic Imaging, College of Health Science, University of Sharjah, Sharjah, United Arab Emirates
[2]Research Institute for Medical and Health Sciences, University of Sharjah, Sharjah, United Arab Emirates
[3]Operational Research Center in Healthcare, Near East University, Nicosia/TRNC, Mersin 10, Turkey
[4]Department of Biomedical Engineering, Near East University, Nicosia/TRNC, Mersin 10, Turkey
[5]Department of Biochemistry, Faculty of Basic Medical Sciences, University of Jos, Jos, Nigeria
[6]Applied Artificial Intelligence Research Center, Department of Computer Engineering, Near East University, Nicosia/TRNC, Mersin 10, Turkey
[7]Department of Radiology, Brain Health Imaging Institute, Weill Cornell Medicine, New York, NY, United States

Contents

20.1 Introduction

Cardiac resynchronization therapy (CRT) has been shown to enhance the quality of life, exercise ability, and outcome of NYHA functional class III–IV, left ventricular ejection fraction (LVEF)-0.35, and QRS duration of around 120 Ms in patients with heart failure. Following the publication of randomized prospective clinical trials, the indication for CRT therapy has been continuously extended. However, it has not been

Practical Design and Applications of Medical Devices.
DOI: https://doi.org/10.1016/B978-0-443-14133-1.00016-1

prospectively evaluated, among other important unanswered questions about CRT, whether patients with previously implanted right ventricular pacemakers or cardioverter defibrillator (ICD) systems are affected compared with patients undergoing de novo CRT implantation they reap similar benefits from resynchronization therapy. In the core, it is roughly placed right underneath the sternum. It is found in the thorax on top of the diaphragm [1]. The heart is in the cavity that is called the mediastinum between the lungs. The human heart is divided into four chambers; the left and right atrium and the left and right ventricle. It is responsible for pumping blood from the heart to the rest of the body. The right atrium receives deoxygenated blood from the body through the vessels and pumps it to the right ventricle through the tricuspid valve, and the right ventricle pumps the deoxygenated blood to the lungs to get oxygenated. Then the left ventricle receives the oxygenated blood through the vena cava, and then it pumps the blood to the left ventricle through the mitral valve. The left ventricle then pumps the oxygenated blood to the rest of the body. The process of pumping blood to and from the body is called the circulatory system. The body needs this oxygenated blood because this is how nutrients are provided to the rest of the body, and in exchange, the blood also carries all the unwanted products, e.g., carbon dioxide, and it is sent through the right side of the heart, where the right atrium sends the blood to the right ventricle and the blood is pumped into the lungs, this is where the carbon dioxide is released from the body when we exhale [1−4]. The blood delivers oxygen to the body cells. The body needs healthy oxygenated cells for a person to stay alive. If the oxygen-rich blood does not circulate the way it is supposed to, the person may die. The heart is able to send oxygenated blood to the veins and arterioles. The blood vessels that carry blood away from the heart are called arteries, and the blood vessels that carry blood to the heart are called veins. When a heart is in a healthy state, it makes a lub-dub sound after each beat. This sound comes from the heart valves shutting off blood inside the heart. The first sound, called the lub sound, occurs when the mitral and tricuspid valves are closed [5]. The second sound, called the dub sound, occurs when the aortic and pulmonary valves are closed, i.e., after the blood has been pumped out of the heart. The heart is able to pump 7200 L of blood per day. The average male heart weighs around 280−340 g. In females, it weighs around 230−280 g. An adult heart beats about 60−80 times per minute, and a newborn baby's heart beats faster than an adult heart, which is about 70−190 beats per minute. During blood circulation, the nutrients collected from the small intestine are pumped to provide nutrients to the cells in the body. The circulatory system also carries all the waste products in the body, for example, nitrogenous waste and salts, from the cells to the kidney, where the body releases them as waste. For the heart to pump this blood, it needs current, an electrical system that is able to send electricity to the left and right ventricles, to pump the blood. This electrical system keeps the heart beating at a normal rate, and when there is a problem with this electrical system, the heart experiences something called arrhythmia. Arrhythmia is caused by the abnormal beating of the heart; when the heartbeat is too fast, it is called tachycardia, and when the heartbeat is too slow, it is called bradycardia [4]. The human heart is divided into three different layers called the epicardium, myocardium, and endocardium.

Heart muscles are divided into four distinct styles of chambers. These are the left atrium, right atrium, left ventricle, and right ventricle. The thin wall is composed of the left and right atrium, while the thick muscle is composed of the left and right ventricles [6−9]. The heart consists of the pulmonary valve, tricuspid valve, aortic valve, and mitral valve.

20.1.1 Materials/Components

20.1.1.1 Microchip 18F4520 PIC Processor

Microchip 18F4520 PIC (programmable intelligent computer) Processor was used in making the remote monitoring pacemaker. It is derived from PIC1650, and it came from a family of microcontrollers made of microchip controllers. The benefit of all PIC18 microcontrollers with high computational efficiency is high endurance, enhancing flash program memory. The PIC18F4520 has the ability to reduce the battery consumption of the device during performance by a number of ways:

1. The alternate run mode helps to clock, during code execution; 90% of power consumption can be reduced by the controller from the timer1 source or the internal oscillator block.
2. The controller can also run with its core CPU disabled with several idle modes, but the peripherals are still working. This can help to reduce the amount of power used to as less as 4% of the operation requirement.

20.1.1.2 Multiple oscillators

The material used for the oscillator is crystal or ceramic, with four crystal modes. There are two external clock modes, providing the possibility of two pins or one pin being used. The external clock mode is fitted with two external RC oscillator modes with the same pin.

20.1.1.2.1 Important features of multiple oscillators

Self-programmability Under internal software control, these devices can generate their own program memory space. This is achieved by using a bootloader routine that can be stored on top of the program memory at the boot block.

10-bit A/D converter This module is capable of integrating a programmable time of acquisition that allows the selection of a channel and the start of conversion without waiting for a sampling cycle.

20.1.1.3 Microchip PICkit 2 and PICkit 3

This appliance comes from a family of PIC controller programmers. It is used for programming microcontrollers and debugging them. The features of PICkit 1 include eight LEDs, a switch, and a potentiometer. The default program rotates the serial LEDs. With a bottom and potentiometer on the PICkit board, the direction and speed of rotation of the light displays can be changed. The distinction between PICkit 1 and PICkit 2 is that PICkit 2 has its own debugger device that attaches the programmable chip to be monitored, while PICkit 1 is a single unit.

20.1.1.4 Model USB 6008

It connects to eight channels of analog input, two channels of analog output, 12 channels of digital input/output, and a 32-bit counter with a USB interface at maximum speed. The safety precautions to be taken when using this device are that you can use a dry cloth if you need to clean this device. Until re-use, the system should be fully dry and free from pollutants. When operating the unit it is important to make sure that the operating level is at or below grade 2. Grade 1 means no pollution, grade 2 means that in most situations, there is only nonconductive pollution, and grade 3 means that there is conductive pollution.

20.1.1.5 Accelerometer

This is the unit used for proper acceleration calculation. This is an electronic sensor capable of calculating an object's location in space, and it also accesses the object's movement. The acceleration is used to calculate an object's velocity. The acceleration tests the gravitational pull while the object is at rest, which is g = 9.81 m/s^2. If the object's acceleration is in free fall, it can measure up to zero. There are many applications for accelerometers in inertial navigation systems for aircraft and missiles, including highly sensitive accelerometers. In addition to resolving devices, they are also able to track vibrations. In tablet computers and digital cameras, accelerometers are often used so that the images on the display can be viewed upright. There are two types of forces for acceleration: static and dynamic forces. Static forces, such as gravity and friction, are forces that are continually applied to the material. Although dynamic forces are moving forces, they are applied at different rates to the object. There are three different types of accelerators: piezoresistive, piezoelectric, and capacitive accelerometers. The piezoelectric accelerometer has the ability to detect changes in acceleration using the piezoelectric effect. The piezoresistance accelerometer is capable of increasing its resistance to the amount of pressure applied to it. The most utilized form is the capacitive kind, which uses electrical capacitance shift to calculate objects' acceleration.

20.1.1.6 Lithium battery

This is a rechargeable battery that can be used in military and aerospace applications for portable electronics and electric vehicles, which are also becoming very popular. Nickel-cadmium has long been the only appropriate battery for portable equipment. It was only in the 1990s that lithium ions emerged, and now they are the fastest growing and most promising battery. Lithium has the greatest electrochemical potential and the greatest energy density for its weight, being the lightest of all elements. Another benefit of lithium batteries is that for most of today's mobile phones, a single-cell battery is used, and the lithium ions have a high cell voltage of 3.6 V, allowing for just one cell battery pack capacity. During discharge, lithium ions pass from the negative electrode through the electrolyte to the positive electrode. The external electrical power source applies an overvoltage while charging, causing a charge current to flow from the positive to the negative electrode inside the battery. Although they have many benefits, lithium-ion batteries have certain limitations. To ensure safe operation, it is fragile and requires a safety circuit; on most packs, the maximum charge and discharge current is limited to between 1 C and 2 C. It has low maintenance. One of the problems faced is the deterioration of the batteries over time.

20.1.1.7 Sensing amplifier

The sensing amplifier is a circuit used to detect low-power signals that represent a data bit (1 to 0) stored in a memory cell when data is read from the memory and to amplify the small voltage swing to an identifiable logic level so that the data can be interpreted correctly by logic outside the memory. The benefits of sensing amplifiers are real-time overcurrent protection, current and power monitoring for system optimization, and current measurement for closed-loop feedback. Modern sense amplifiers that are used these days have 2 to 6 transistors, while in the earlier days, the sense amplifier had up to 13 transistors. There are two different types of circuitries for the sense amplifier, differential (voltage) and non-differential (current).

20.1.1.8 Pacing pulse generator

This is an electronic circuit that is used to generate a rectangular pulse. One of the important uses of pulse generators is in digital circuits, but they are also used in analog circuits. In biotechnology and industrial applications, pulse generators are commonly used, such as ultrasound imaging, biofouling prevention, bacterial decontamination, transmission of genes and medications, air clearing, etc. This is the unit that provides the desired electric pulse field in the pacemaker. Pulse generators require a high DC voltage to power it. This means that additional high DC voltage is needed in the circuit design.

20.1.1.9 Comparator

A comparator circuit regulates the transistors, and when the maximum cell voltage set by the comparator is reached, it proportionally bypasses the current around the cell. This is a cell that compares and outputs a digital signal with two voltages and currents that indicate which one is bigger. They consist of two signals, one of which comes from the output and one from the input.

20.1.1.10 Filter

A filter is capable of not moving certain frequencies while attenuating those frequencies at the same time. A filter can extract from signals important frequencies that also contain unwanted frequencies. There are four different filter types: low pass filter, high pass filter, band pass filter, and notch filter. A frequency range of $100-300$ Mhz is most sensitive to passive filters. Ideally, a filter will not add new frequencies to the input signal, nor will it change the signal's component frequencies, but only after determining the relative amplitudes of the different components of the frequency and their relationship with the phase.

20.1.2 Working principle

The sinoatrial node is the heart's electrical system capable of providing current to the heart. The sinoatrial node, which is located in the left atrium, sends an electrical impulse to the AV node (atrioventricular node), and the AV node passes the electrical impulse into the muscle tissue of the left and right ventricles. The left and right ventricles contract, and blood is pumped into the body. In some cases when a patient experiences arrhythmia, the SA node may not send electrical impulses properly [10,11]. The electrical impulse is either too fast or too slow. When this happens, the patient experiences shortness of breath, dizziness, palpitations, and fainting. An artificial electrical system has been designed to help patients with such conditions. This device is called a pacemaker.

The pacemaker was invented over 50 years ago, but the first successful pacemaker that was recognized by people was created in 1926 by Dr. Albert Hyman [6,10]. Since then, the pacemaker has become more sophisticated and a lot smaller than it was back then. Pacemakers consist of a battery, pulse generator, and lead. However, lead is one of the most common factors of pacemaker malfunctions. Recent pacemakers are made with leadless devices. Although pacemakers are used in stimulating the heart, recent pacemakers are also capable of sensing the activities of the heart and are also wireless. Wireless pacemakers contain a remote monitoring system that allows the clinician to monitor the patient from afar. Remote monitoring systems provide access to useful data, including the identification of issues associated with arrhythmias and heart failure. Wireless pacemakers use an electrode that is implanted into the heart muscle. A lot of people around the world today are dying of heart problems, and pacemakers have been able to provide so many patients with second chances and save millions of lives. Especially with rate-responsive pacemakers, it is possible to detect the physical and emotional conditions of the patient and adjust to the condition or activity the person is doing. Rate-responsive pacemakers have helped patients get back to their normal lives while they are being monitored. Transtelephonic patient monitoring has been used for decades, it includes a lot of patient revisits. Every 3–12 months the patients receive checkups because these are the times when either the battery needs replacement or the pacemaker needs better calibration. Most recent pacemakers have transmitters that enable the physician to download the data, and analyze the health of the patient without constant patient revisits that involve so much gruesome work [4]. Therefore a remote monitoring system is able to provide patients with quality health care and is a lot more safer. There are different types of remote patient monitoring systems around the world, for example, the Biotronik Home Monitoring, CareLink Network, LATITUDE, etc. The different types of pacemakers include single-chamber pacemakers, biventricular chamber pacemakers, and dual-chamber pacemakers. Although pacemakers have been able to improve the lives of so many people, they also have their limitations.

20.2 Method

One of the leads is placed in the right atrium, and the other is placed in the right ventricle. A dual-chamber pacemaker uses two leads. This system is capable of monitoring both the right atrium's and the right ventricle's electrical activity to check if pacing is necessary. The pacing pulse is built in a way that can mimic the natural movement of the heart when pacing is required. To regulate the pacing logic, the 18F4520 PIC processor microchip is used. A cardiologist can configure current pacemakers to choose appropriate pacing modes for individual patients. There are situations where a combination of both pacemakers and a defibrillator is used by the cardiologist in a single implantable unit [11]. The PICkit 2 or PICkit 3 microchip can be used as a pacemaker board to program and debug the PIC18F4520 chip. It is possible to connect both of these devices to any machine through any accessible USB port. This is the kind of pacemaker that resembles the natural beating of the heart. There is an accelerometer on the pacemaker board that is used to track activity and facilitate the creation of rate-responsive software. The pacemaker generator is largely sealed with a lithium-ion battery, a sensing amplifier that processes the electrical manifestation of heartbeats that occur spontaneously as sensed by the heart electrode, a pacing logic for the pacemaker, and an output circuit that provides the electrodes with the pacing impulse. A 9-volt battery can power this pacemaker, but it is more convenient to power it using an external AC adapter. When, within a standard beat-to-beat time frame, the pacemaker fails to detect a heartbeat within a normal beat, it will activate a heart ventricle with a quick low-voltage pulse. On a beat-to-beat basis, the method of detecting and stimulating the heart will begin. The most advanced pacemaker is the pacing mode of the basic ventricular "on demand" with automatic rate adjustment for exercise, which is sufficient when no synchronization with the arterial beat is required. Present pacemakers with thousands of rows of codes are highly advanced cardiac rhythm operators. These devices are able to correct a variety of complex anomalies in the heart. To accommodate changes in the heart as it ages, these devices can also be easily reprogrammed. It was claimed that the original hardware sensing circuit had noise immunity and the ability to sense down to 37 uV [7]. This was accomplished by offering, under the guidance of a microprocessor, a variable gain and reference level for the circuit. The pacing circuit is capable of supplying selectable voltage amplitudes ranging from 1.2 to 7 V almost continuously, and the pulse width is limited only by the maximum lock frequency. The lead impedance measurement circuit has the capacity to provide less than 1% error for an accurate impedance measurement. The low-power design approach plays a key role in the design, since most embedded portable devices are battery operated. A pacemaker is a real-time system operated by a computer with predefined task priorities. The microcontrollers that should be used should have a low consumption of power, and memory space is also needed. The PIC18F4520 microchip is widely used in many pacemakers used today. It has low power consumption and up to 16 bits of memory space, as well as a high speed of 40 Mhz. The life cycle of the battery, its size, width, and the materials it is made of should all be considered. One of the most important considerations is that it should last for 10−12 years. The front end can sense the voltage produced by the heart's pumping behavior, which is small and carries a lot of noise. There are lots of components in the circuit, such as the amplifier, filter, level shifter, synchronization circuit, etc. A simple pacemaker circuit is demonstrated in Fig. 20.1. A pulse of 5−7.5 V, a multiplier, and a transfer network are used to speed an erratic heartbeat. The pacemaker has an external comparator. The signal is often detected by unipolar or bipolar electrodes, and a low-noise preamplifier amplifies it. It goes through the filter when the lead electrodes are preamplified and the

Figure 20.1 Pacemaker circuit.

signal is amplified again. After this the signal in the pulse generator is synchronized, and the pulse generator processes it. Another pulse generator goes through it, and then the signal is sent to the heart. Typically, pacemaker software is equipped with timed automata [9].

20.3 Results

The model of the pacemaker explains the behavior of the software used in modern ICPs with dual chambers. The network consists of five automatons, each of which handles a particular feature of the cardiac cycle: the lower rate interval, the upper rate interval, the atrioventricular interval, the auricular postventricular interval, the refractory period, and the refractory ventricular period. The condition of the scheme is decided by the union of the states of the parts. The PVS language has been translated into this model. Using our PVSio-web tool, which provides a communication infrastructure to enable the exchange of simulation events between PVS and MathWorks/Simulink, the translated model was interconnected to the ICP-web service, where the proposed system recorded a great performance. As shown by Cleland et al. [10], 37/56 de novo implanted patients (66%) and 10/17 upgraded patients (59%) were considered responders to CRT (P 1/4 0.80) according to the predefined criteria. The NYHA functional class, LVEF, and NT-proBNP levels showed a substantial change in responders relative to nonresponder patients as part of the combined concept. The only complications to be listed were one pneumothorax with total left lung collapse requiring drainage and resolving without sequels, and one thrombosis in the posterolateral branch where the electrode was manipulated (incidentally discovered during control venography) that fully developed asymptomatic medicine. Left ventricular electrode dislocation occurred in three patients (14%), with loss of left ventricular capture in one patient and an excessive rise in the pacing threshold in the other two patients. Both had the electrode repositioned successfully in two cases, while the third case, without diaphragmatic stimulation, had no current unipolar pacing thresholds lower than 3.7 V. This made chronic pacing impossible with our generator in pseudobipolar mode. CRT de novo implantation was generally effective in 90%–95% of patients in recent, controlled clinical trials [9]. By comparison, in a study of 56 patients with CRT upgrade procedures, implantation performance was recorded in only 82% of the attempts [10]. Nearly all other CRT upgrade studies were retrospective in nature and only included patients with successful implantation or with no success or failure registered. Our analysis shows that CRT can be upgraded. Accomplished with equal success as implantations de novo, in upgrade procedures, complications such as more complicated access from the right side to the coronary sinus, the movement of preexisting chronically inserted leads that may have developed into the venous wall, or may hinder the subclavian veins, or may be connected to the tricuspid valve, preventing the cannulation of the coronary sinus ostium, should be expected. Therefore it may be advisable to perform a subclavian vein angiography before planning a CRT upgrade procedure, particularly in patients who already have several transvenous leads, to remove a full subclavian or brachiocephalic vein obstruction [11]. Our study also shows that the upgrade times for implantation and fluoroscopy are comparable to those of fluoroscopy. Lastly, in both patient classes, there were no major variations in CRT-related complications.

20.3.1 Discussion

We have identified a device that is capable of assisting a cardiologist in reprogramming an implanted pacemaker. We used systematic and logical techniques in developing the method. As a result it is possible to explain the underlying methods of representation and reasoning correctly and prove the system's properties. Technological advances have generalized the CRT technique by using minimally invasive procedures that allow the selection of the left ventricular area to be paced. Our investigation reveals recent series that achieved left ventricular pacing of 84%–92%.

We must note that some procedures were time-consuming; in fact, 40% of our cases showed longer ventricular pacing. We suggest using a radioscopy device with appropriate resolution for viewing 0.014 guides and integrated image storage and comparison functions to achieve high success rates, without time as a limiting factor. The use of traditional electrophysiology may also promote coronary sinus catheterization. Using steerable catheters and recording the intracavitary electrocardiogram in electrocatheters or in particular difficult situations.

Failure to implant a left ventricular electrode, which chronologically corresponded to one of the last patients in our series, was probably triggered by the former. Another case in which continuous left ventricular pacing was not achieved showed a unipolar intraoperative threshold of 3.7 V at 0.5 Ms, while capture with a generator programmed in pseudobipolar mode at full energy production was unlikely (7.5 V 1.0 Ms) [11]. Other authors have already stated this fact, and a generator with a separate output for each ventricle, 19 or even bipolar electrodes, may have solved it [9,10]. More time was required for implantation in patients with ischemic heart disease, and their left ventricular thresholds were higher. This presents significant technological problems, which are likely to be triggered by existing myocardial regions that are necrotic and unexcitable. Since the design dimensions of our study and the subjective nature of some of the clinical variables examined (functional class, number of admissions) are relatively small, it cannot be inferred if the clinical benefits found were due to therapy. Nevertheless, the information collected for our series is consistent with the results of most multicentric studies that show similar improvements in functional class and quality of life and fewer admissions. Another development appears to be the higher oxygen uptake associated with BIV pacing showing up in postimplantation acute studies, with the requisite limitations of these acute tests. No progress was seen in three patients (14%). Ventricular dyssynchrony makes it possible to refine the enhancement prediction criteria. A further contribution to the correctness of the method is extracted from the use of model-based approaches in the creation of the knowledge base for the system. For this reason it seemed necessary for the reprogramming of a pacemaker to use causal models of abnormal behavior as a basis for domain knowledge behavior. It is interesting that the bipolar stimulation test differs from the simplification test. For the proposed model, the hypothesis was held. Both the anode and the cathode lie at the tip of the pacemaker lead, so that equal and opposite currents flow through the lead wires, theoretically providing the magnetic induction field concatenated with the pick-up field with an almost null overall contribution. The remaining non-null contribution is that of the ionic currents generated in the saline solution by the two electrodes, which are localized around the tip of the lead of the pacemaker, thus showing a weaker magnetic coupling with the pick-up coil as it is more distant from it. Nevertheless, with bipolar stimulation, the sensor was able to detect the pacemaker pulses. Because of the edges in bipolar stimulation, pacemaker pulses are much more rapid than in unipolar stimulation. Any inherent limitations are exposed by the sensor. Currently, the reaction of the coil to every electromagnetic impulse is a damped sine curve with the same natural frequency and damping ratio. It has been observed that sources of intense electromagnetic interference can corrupt stimulation rate measurements by triggering sufficiently high amplitude coil responses to be detected at short distances from the sensor. Our experience is that standard models are not necessary for the advice and diagnosis of suitable pacemaker settings, health conditions, and pacemaker issues. The device only covers half of the whole domain at present. The method will be expanded in the near future by adding other parts of the issue domain as well. In addition, a program that can be incorporated with the pacemaker programmer software will replace the current implementation, so that the user can consider the device a functional addition to the programmer software. Integration also has the advantage that it is not required to enter pacemaker data by hand, but can be automatically extracted from the pacemaker system. The ability to view rate histograms provides other benefits to remote control of pacemakers. Rate regulation is important in patients with established atrial fibrillation to prevent tachycardia-mediated cardiomyopathy and recurrent remote cardiomyopathy [4]. Access to rate histograms can allow clinicians to track the progress of rate control more effectively. In addition, in recent years, the possible adverse effects of right ventricular pacing have gained popularity. In patients who had over 40% ventricular pacing, a substudy from MOST indicated an increased risk of hospitalization for heart failure. With right ventricular pacing, the incidence of atrial fibrillation has also increased. The percentage of right ventricular pacing given is now recorded by pacemakers, thus enabling clinicians to make programming adjustments to restrict the amount of right ventricular pacing. High ventricular rate episodes that have been reported by nonsustained ventricular tachycardia, which may be an indicator of cardiomyopathy in patients not previously suspected of systemic heart disease, can also be indicated by pacemakers. It is then possible to perform adequate screening to determine the patients' risks of sudden cardiac death and the potential for an ICD upgrade. Remote control devices are commonly used to track ICDs, but have not yet been studied for pacemaker monitoring. To our understanding, this is the first big, multicentered trial comparing the usefulness of remote-control systems with the latest targeted temperature management (TTM) and office visit systems. Our theory is that the pacemaker is clinically important for arrhythmic problems that patients may acquire, that pacemaker diagnostic data are complex and multiple, and that repeated access to these data through remote monitoring can lead to earlier clinical acts.

20.4 Conclusion

There is a rise in both complexity and functionality as far as biomedical electronic systems are concerned. To incorporate usable and secure systems, it is important to understand both the human body and the embedded computer applications. This structure is a step in this direction, bridging the gap between biology, computer science, and electrical and computer engineering. We defined a relatively inexpensive, extensible framework for which the design of new academic pacemakers and the study of current prototypes may be used. The purpose of this framework is to provide a convenient forum that can be used in real-time embedded device design and verification for research and teaching. To design a system, no matter what its intent, it is important to study the field of intended use to gain sufficient information to create a comprehensive system. Some advanced features, such as telemetry, optical signal processing, and power control, are included in the pacemaker specifications.

References

[1] S. Mittal, M.,E. Kloosterman, et al., How Wireless Remote Monitoring Improves Clinical Benefits, The Arrhythmia Institute at Valley Hospital, Ridgewood, NJ and New York, 2018.
[2] Market Report, World medical devices market. Acmite Market intelligence, 2014.
[3] V. Sastri, Plastics in medical devices: properties, requirements and applications, 2013.
[4] D. Halperin, T.S. Heydt-Benjamin, B. Ransford, S.S. Clark, B. Defend, W. Morgan, et al., Pacemakers and implantable cardiac defibrillators: software radio attacks and zero-power defenses, 2008 IEEE Symposium on Security and Privacy (sp 2008), IEEE, 2008, pp. 129–142.

[5] WHO, 2010. Barriers to innovation in the field of medical devices: background paper 6, August 2010 (No. WHO/HSS/EHT/DIM/10.6). World Health Organization.

[6] N. Varma, J.P. Piccini, J. Snell, A. Fisher, N. Dalal, S. Mittal, Relationship between level of adherence to automatic wireless remote monitoring and survival in pacemaker and defibrillator patients, Journal of the American College of Cardiology 65 (24) (2015) 2601−2610.

[7] C. Nixon, J. Smith, T. Ulrich, R. Davis, C. Larson, K. Cha, Academic dual chamber pacemaker, University of Minnesota, Final Report, 2007.

[8] <http://www.circuito.io/app> Retrieved March 24, 2021.

[9] A. Rajeef, D. David, et al., A theory of timed automata, Theoretical Computer Science 126 (2) (1994) 183−235.

[10] J.G. Cleland, J.C. Daubert, E. Erdmann, N. Freemantle, D. Gras, L. Kappenberger, et al., Cardiac Resynchronization-Heart Failure (CARE-HF) Study Investigators: the effect of cardiac resynchronization on morbidity and mortalityin heart failure, The New England Journal of Medicine 352 (2005) 1539−1549.

[11] C. Leclercq, S. Cazeau, D. Lellouche, F. Fossati, F. Anselme, J.M. Davy, et al., Upgrading from single chamber right ventricular to biventricular pacing in permanently paced patients with worsening heart failure: the RD-CHF study, Pacing and Clinical Electrophysiology 30 (2007) S23−S30.

CHAPTER TWENTY ONE

The design of a noninvasive blood pressure measurement device

Dilber Uzun Ozsahin[1,2,3], **Declan Ikechukwu Emegano**[3,4], **Belal J.N. Abuamsha**[4], **Basil Bartholomew Duwa**[3,4] and **Ilker Ozsahin**[3,5]

[1]Department of Medical Diagnostic Imaging, College of Health Science, University of Sharjah, Sharjah, United Arab Emirates
[2]Research Institute for Medical and Health Sciences, University of Sharjah, Sharjah, United Arab Emirates
[3]Operational Research Center in Healthcare, Near East University, Nicosia/TRNC, Mersin 10, Turkey
[4]Department of Biomedical Engineering, Near East University, Nicosia/TRNC, Mersin 10, Turkey
[5]Department of Radiology, Brain Health Imaging Institute, Weill Cornell Medicine, New York, NY, United States

Contents

21.1 Introduction

Blood pressure is the force that pushes the blood against the vessels [1]. It is measured in millimeters of mercury (mmHg). Blood pressure is designated and written in two forms, systolic and diastolic when the heart rests between the beats. Normal pressure is given as 120/80 mmhg, i.e., any systolic value less than 120 and lower than 80 [2]. Blood pressure determines a person's circulatory status, an indicator of irregular pressures,

Practical Design and Applications of Medical Devices.
DOI: https://doi.org/10.1016/B978-0-443-14133-1.00004-5

i.e., hypertension and hypotension. Hypotension causes an insufficient quantity of oxygenated blood to reach the brain, and this condition can lead to distress and syncope. Monitoring blood pressure accurately requires the use of properly regulated certified monitoring devices. Good blood pressure monitoring is complicated by irregular heart conditions [3]. Realistic simulations could be utilized to check and calibrate wireless blood pressure sensors as well as provide blood pressure measuring instructions. Many hand simulations are already widely viable [3]. The majority of such simulations help you take your blood pressure sensor using the sphygmomanometer approach, and some also allow you to simulate Korotkoff noise using the palpatory technique [4]. Such equipment is often expensive and limited in its ability to set simulated variables.

Blood pressure measurement is essential for diagnosing a variety of health issues. In the past, the traditional methodology of blood pressure measurement was by listening for Korotkoff sounds from a cuff using a mercury (oscillometric) sphygmomanometer [5]. This is regarded as the reference standard in blood pressure evaluations. Although, most of the concern about cuff pressure is reliability and withstanding clinical stresses. As a result, the development of automated devices became in vogue. The quasi pulse pressure noninvasive blood pressure (NIBP) monitoring [6] was the first to be developed. They took blood pressure repetitively and, in most cases, were attached to an alarm bell for vital sign parameters. These pressure devices are not cost-effective. The next development was the blood pressure spot check device. They are used in single measurements and routinely. The ambulatory NIBP measuring device is used for a predefined interval. The most used at home is the automatic self-check blood pressure device. It is very cost-effective and most clinicians prefer using it in clinics. However, users of these devices need to be mindful of the restrictions that come with them. Automatic blood pressure machines come in four types [7]: (1) NIBP monitors that cycle through automatic testing cycles. These carry out repeated measurements at predetermined intervals of time and frequently include alerts for vital sign indicator monitoring. They are cost-effective for monitoring patients at the bedside in clinical settings. (2) NIBP spot-check monitors are also designed for routine clinical assessment. (3) Another type of automatic blood pressure is the NIBP ambulatory device for blood pressure that records patient data over 24 hours. (4) Finally, the NIBP spot-check monitoring devices were originally meant for home use, but nowadays clinicians make use of them in the clinic because of their low cost [8].

21.1.1 Blood pressure measuring equipment

The blood pressure device could be used as a manual auscultatory procedure or an automatic oscillometer methodology. The auscultatory technique uses an inflated forearm cuff to block the brachial aortic valve. The cuff is then gradually deflated while the Korotkoff sounds are heard through a stethoscope. The measurement on the sphygmomanometer is the patient's systolic and diastolic blood pressure. The mercury, aneroid, and electronic sphygmomanometer are subclasses of the auscultatory manual technique. The oscillometric method is used by most noninvasive automated devices and they include the automatic soft check, devices worn on the risk, and finger. Others are the NIBP used at the spot, automated cycling NIBP, multiparameter, and ambulatory subclasses of blood pressure measurement method.

21.1.2 Physiology of heart and circulatory cascade

The heartbeat and circulatory vessels that pump blood provide nutrition and oxygenation to the tissues. The cardiac pumps, also known as circulatory pumps, are made up of four sections: left atrium, left ventricle, right atrium, and right ventricle [9].

The right atrium is filled with oxygenated blood, and the cardiac vein is indeed drained into the right atrium. The oxygenated blood reaches the left atrium through the pulmonary vein and can then be forced further into the left ventricle via the tricuspid valve by the activation of the atrial musculature. When the pressure in the chamber rises high enough, the aortic valve opens, allowing blood to flow into the aorta and circulate throughout the body [10]. With each heartbeat, the cycle of blood circulation is resumed. However, the heart has two significant beats: systolic and diastolic. The systolic phase is responsible for the ventricular pump. The diastolic stage occurs when the heart chambers dilate as the blood is pumped [11]. The other various beats are ventricular diastole and systole, atrial diastole, and systole. The ventricular diastole causes the depletion of oxygenated blood away from the ventricles and then to the pulmonary arteries. Atrial diastole has oxygenated blood arising from the pulmonary vein entering the left atrium and the vena cava filling the atrium [12,13].

21.1.2.1 Tonometry technique
The tonometry technique was developed during the 20th century according to oscillometric and auscultatory principles [14].

Depending mostly on temporary blockage of the arterial (blood flow is never completely obstructed), these approaches permit the continual recording of arterial blood pressure waves at the expense of precise positioning of both the sensor devices.

Pressman and Newgard developed noninvasive tonometry in 1963 [15,16], motivated mostly by the pioneer innovations of Vierordt and Marey through their earlier research in visual measurement. This technique involves exerting pressure on the middle of a peripheral artery (over an underlying bone) until the artery begins to collapse. Pressman and Newgard [17] established this using the linearity model of a measuring apparatus. They further explained that the vertical distortions seen by the sensor were proportionate to arterial blood pressure.

By maintaining the continuous position of such tonometric arterial bone systems, the tonometer converts the detected vertically displaced signals into an electrical impulse representing arterial blood pressure waves. Drzewiecki created more advanced models describing the physics of applanation tonometry observations in the 1980s [17]. The brachial aorta is employed as a selected location in this procedure. The circumferential pressure of the arteries is measured using a cylinder network of strength detectors. The arterial bone system is linked to every pressure detector. This technology is painless and could be utilized in monitoring the heart rate.

21.2 Literature study

The implementation and design of cuffless NIBP from the arterial region, is among the most prevalent method used to design a noninvasive blood pressure measurement device. This device measures values of high blood pressure, a diseases in the world that can lead to a stroke is hypertension, sometimes referred to as elevated blood pressure. Continuous monitoring of the patient's blood pressure readings is essential for the accurate diagnosis of hypertension. Within the scope of this investigation is their evaluation of a specialized cuff-less monitoring device that calculates blood pressure levels alone without the requirement of calibration. The study uses continuous measurements from 40 healthy individuals about 20 to 30 years of age. There were 30 males and 10 girls in the group. Our measuring procedure included 15 minutes of continuous electrocardiography

(ECG) and photoplethysmography (PPG) in each of the three sessions, which were defined as restoration, rest, and cycling activity. The systole and diastolic pressure estimations were done using up to nine characteristics that were picked from the 34 potential features that were collected from ECG and PPG signals. To get the most accurate findings possible, we validated them using three different predictive models, including regression analysis, regression using support vector machines, and regression using multilayer perceptron. The research offers a potentially useful method for the development of cuffless blood pressure monitoring systems of the present day. CAS Medical System in their paper explains that NIBP is a crucial indicator that is frequently used for assessing cardiovascular health. The NIBP evaluations that are obtained by vital signs monitors are accomplished by the incorporation of available commercially OEM blood pressure module technology into the monitors themselves. The oscillometric method is commonly utilized by these devices to perform the task of determining the patient's blood pressure. The conformity of NIBP devices with standards, such as the Association for the Advancement of Medical Instrumentation (AAMI)/American National Standards Institute (ANSI) SP10−2002, is verified before their approval for usage in clinical settings [18].

21.3 Mechanism of pulse oximeter

Pulse oximetry offers a new revolution for the continual, precise, and quasi-monitoring of oxygenation. Regardless of its widespread use, numerous suppliers are unaware of its fundamental operational principles. This information provides the critical framework for recognizing the limitations of pulse oximeters and detecting when their values are inaccurate [19]. Pulse oximetry is common in healthcare. It is frequently considered the fifth vital sign [20]. The two underlying concepts of pulse oximetry are (1) the distinguishing factor of oxyhemoglobin (O2Hb) from deoxyhemoglobin (HHb), and (2) methodologies involved in the calculation of SpO2 from the arteriolar chamber of blood [19].

21.3.1 Design approach

A wearable device of nonintrusive consistent blood pressure monitoring is still being researched. PPG, artery tonometry, and pulse wave transit time

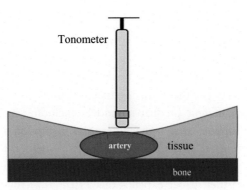

Figure 21.1 Tonometry technique.

(PWTT) approaches have all been extensively explored as measurement methodologies that are attached to the fingers, as can be seen in Fig. 21.1 [21]. The different simulations of the fingers and the apparatus were done using a computer system. The PPG technique is always embedded with a paired photodiode, which has detectors and a sensor that is regularly utilized for artery tonometry. It also comes with fluid cartridge architectures, acoustic patched systems, and a device sphygmomanometer thumb approach are just a few examples of good approaches using new techniques.

The sensor component was combined with a wearable blood pressure monitor to illustrate its blood pressure measurement capabilities in daily usage. Modeling and experimentation were used to carefully verify the equipment's functioning. As illustrated in Fig. 21.1, the brachial artery blood pressure measuring equipment consists of wrist-worn equipment with a mobile device arterial blood pressure monitoring app. The information is sent through Bluetooth Low Energy through both the wearable devices and the monitoring app, eliminating the requirement for a large information-gathering equipment and answering the demand for constant blood pressure monitoring mobility in everyday life. The product's monitoring methodology is mostly built on artery tonometry, with PWTT measurements for good effect. Blood pressure hand simulations are used to evaluate and calibrate quasi blood pressure devices by simulating the hand's movement throughout the monitoring procedure. Many such simulations, such as the simulation of variable artery blood pressure, have been published [21]. Also, the stimulation of the cyclic impulses is used in the calibration.

21.3.2 Methodology of operation

The blood pressure mounting device contains an LCD and is connected to a circuit board, as can be seen in Fig. 21.2. The circuit board has a pump, solenoid, LCD, speaker, pressure sensor, and computer soldering buttons. In the prototype, the pump and the beeper are required. The entire setup is connected to a breadboard. When the circuit is closed the values begin to run pushing air into the device. The setup is easy to use. The finger cup is selected depending on the size of the finger by allowing the patient's finger in the setup. The strap is wound around the arm of the patient in a proximal position, allowing free movement of the patient's finger. A sturdy compound detector was constructed based on the thumb beat in ancient Chinese medicines [22]. It is linked to the exterior via dual tubes upstream, one providing fluid injections and the other for force transference, as can be seen in Fig. 21.2. When a stimulus is detected, the protrusions flex and create compression tension, which may be effectively communicated to the

Figure 21.2 Schematic diagram of blood pressure apparatus in Arduino.

tension sensor component. The tension delivered to the detecting component must be exact, regardless of where the protrusions are squeezed. The primary aim is to make it easier to position the device axially along the wrists. The sensitivity of the censoring device is a result of its thermal polymer sheet, which is both plastic and biocompatible. Diphenyl silicon liquid, which is nonvolatile and biodegradable, is also used.

21.3.3 Simulated outcome on dynamic loads of blood pressure calibration

Following ambulatory blood pressure readings, the baseline blood pressure may be calibrated using the parameters of the transmural pressures. Intravascular tension, administered tension, and hydrostatic tension are all included in the transmural tension. Furthermore, since the hydrostatic tension is equivalent to zero, the transmural tension equals zero. As a result, the starting blood pressure may be determined in one of two ways: when the pulse oscillation intensity hits its greatest or when the PWTT is the biggest. The tension actuator and a gasbag above the sensor device regulate the tension exerted on the wrists by the sensor device. The wrist's tension is then progressively and evenly increased till the captured cardiac pattern began to deform, suggesting that the arterial blood was nearly occluded.

For the 20 individuals, the baseline blood pressure was corrected using the same procedure, The mean blood pressure was calculated from the associated mean tension. The different pressures across systolic and diastolic tension were calculated from the associated oscillation tension since the blood pressure approach according to this research is basically on the tonometry of the artery. The data were translated to systolic and diastolic blood pressures using a 4/6 approach to allow comparisons with commercial cuff sphygmomanometers. The Bland–Altman technique was also used to examine the relationship between both the generated blood pressure and the blood pressure obtained by a commercial sphygmomanometer. The credibility range encompasses all measurement outcomes. Systolic blood pressure has a mean inaccuracy of 4.62 mmHg and a standing inaccuracy of 7.00 mmHg, while diastolic blood pressure has a mean error of 3.00 mmHg and a standard deviation (SD) of 5.10 mmHg.

The findings are consistent with the AAMI guideline of 5–8 mmHg. These aforementioned test findings show that the blood pressure monitoring equipment described in this research is quite stable and adaptable to various persons.

21.4 Conclusions

Artery tonometry is a nonsurgical, high-accuracy technique for measuring blood pressure. Yet, sensor—artery misalignment is a serious issue that prevents artery tonometry from being used. The novelty of this study, of a pulse sensor component, was postulated to measure pulses generated by the heartbeat. In an actual implementation of a wearable blood pressure tracking device, the optimal measurement position is from 2.5 to 2.5 millimeters of the modules, and the pressure variance is within 5.4%, which can be readily done by finger sensing. Different calibration procedures in a practical blood pressure measurement phase may significantly decrease misalignment distortion issues. The device was tested for accuracy and reproducibility on sonography with 20 people, and the process followed the Institute of Electrical and electronics engineers guidelines. The findings show that the functions of the wearable device were effective for blood pressure readings, and the data generated is accurate in accordance with the AAMI objectives. This design is anticipated to continue providing a new option for continual blood pressure tracking in a wearable gadget. However, the design showed scientific knowledge but could not be used to replace the conventional blood pressure apparatus

References

[1] American Heart Association, What is high blood pressure? 2021.
[2] D. Pickering, S. Stevens, How to measure and record blood pressure, Community Eye Health / International Centre for Eye Health 26 (84) (2013) 76. Accessed: Aug. 05, 2022. [Online]. Available: /pmc/articles/PMC3936692/.
[3] B. Ibrahim, R. Jafari, Cuffless blood pressure monitoring from a wristband with calibration-free algorithms for sensing location based on bio-impedance sensor array and autoencoder, Scientific Reports 12 (1) (2022). Available from: https://doi.org/10.1038/S41598-021-03612-1.
[4] K.J. Tietze, Clinical Skills for Pharmacists: A Patient-Focused approach, Elsevier Health Sciences, 2012, p. 200.
[5] K.J. Tietze, Physical assessment skills, Clinical Skills for Pharmacists (2012) 43—85. Available from: https://doi.org/10.1016/B978-0-323-07738-5.10004-3.
[6] W.D. Engle, Definition of normal blood pressure range: the elusive target, Hemodynamics and Cardiology (2012) 49—77. Available from: https://doi.org/10.1016/B978-1-4377-2763-0.00003-2.
[7] The 3 best blood pressure monitors for home use in 2022, Reviews by Wirecutter. https://www.nytimes.com/wirecutter/reviews/best-blood-pressure-monitors-for-home-use/ (accessed August 27, 2022).
[8] B. Zhan, C. Yang, F. Xie, L. Hu, W. Liu, X. Fu, An alignment-free sensing module for noninvasive radial artery blood pressure measurement, Electronics 10 (23) (2021) 2896. Available from: https://doi.org/10.3390/ELECTRONICS10232896.

[9] Structure of the heart, SEER Training. https://training.seer.cancer.gov/anatomy/cardiovascular/heart/structure.html (accessed August 05, 2022).

[10] M.O. Popa, et al., The mechanisms, diagnosis, and management of mitral regurgitation in mitral valve prolapse and hypertrophic cardiomyopathy, Discoveries 4 (2) (2016) e61. Available from: https://doi.org/10.15190/D.2016.8.

[11] T. Nair, Systolic and diastolic blood pressure: do we add or subtract to estimate the blood pressure burden? Hypertension Journal 2 (4) (2016) 221−224. Available from: https://doi.org/10.5005/JP-JOURNALS-10043-0060.

[12] Phases of the cardiac cycle when the heart beats. https://www.thoughtco.com/phases-of-the-cardiac-cycle-anatomy-373240 (accessed August 05, 2022).

[13] P. Reant, et al., Systolic time intervals as simple echocardiographic parameters of left ventricular systolic performance: correlation with ejection fraction and longitudinal two-dimensional strain, European Journal of Echocardiography 11 (10) (2010) 834−844. Available from: https://doi.org/10.1093/EJECHOCARD/JEQ084.

[14] J. Maria Solà Carós, Continuous Non-Invasive Blood Pressure Estimation.

[15] T. Athaya, S. Choi, A review of noninvasive methodologies to estimate the blood pressure waveform, Sensors 22 (10) (2022) 3953. Available from: https://doi.org/10.3390/s22103953.

[16] U.R. Bagal, P.C. Pandey, S.M.M. Naidu, S.P. Hardas, Detection of opening and closing of the aortic valve using impedance cardiography and its validation by echocardiography, Biomedical Physics & Engineering Express 4 (1) (2017) 015012. Available from: https://doi.org/10.1088/2057-1976/AA8BF5.

[17] Development and modeling of arterial applanation tonometry: a review, Semantic Scholar. https://www.semanticscholar.org/paper/Development-and-modelling-of-arterial-applanation-a-Matthys-Verdonck/25a964e53d014cb3b59b0da8-fa68ecccc5583027 (accessed August 06, 2022).

[18] Motion artifact extraction technology for non-invasive blood pressure measurement—medical design briefs. https://www.medicaldesignbriefs.com/component/content/article/mdb/pub/briefs/9502 (accessed August 27, 2022).

[19] E.D. Chan, M.M. Chan, M.M. Chan, Pulse oximetry: understanding its basic principles facilitates appreciation of its limitations, Respiratory Medicine 107 (6) (2013) 789−799. Available from: https://doi.org/10.1016/J.RMED.2013.02.004.

[20] A. Hakemi, J.A. Bender, Understanding pulse oximetry, advantages, and limitations, Home Health Care Management & Practice 17 (5) (2016) 416−418. Available from: 10.1177/1084822305275958.

[21] D. Castaneda, A. Esparza, M. Ghamari, C. Soltanpur, H. Nazeran, A review on wearable photoplethysmography sensors and their potential future applications in health care, International Journal of Biosensors & Bioelectronics 4 (4) (2018) 195. Available from: https://doi.org/10.15406/IJBSBE.2018.04.00125.

[22] A.C.Y. Tang, Review of traditional chinese medicine pulse diagnosis quantification, Complementary Therapies for the Contemporary Healthcare (2012). Available from: https://doi.org/10.5772/50442.

CHAPTER TWENTY TWO

Electromechanical hand-driven electromyogram signal

Dilber Uzun Ozsahin[1,2,3], Declan Ikechukwu Emegano[3,4],
Samer M.Y. Altartoor[4], Mohammad Eyad Osama Yousef[4],
Basil Bartholomew Duwa[3,4] and Ilker Ozsahin[3,5]

[1]Department of Medical Diagnostic Imaging, College of Health Science, University of Sharjah, Sharjah, United Arab Emirates
[2]Research Institute for Medical and Health Sciences, University of Sharjah, Sharjah, United Arab Emirates
[3]Operational Research Center in Healthcare, Near East University, Nicosia/TRNC, Mersin 10, Turkey
[4]Department of Biomedical Engineering, Near East University, Nicosia/TRNC, Mersin 10, Turkey
[5]Department of Radiology, Brain Health Imaging Institute, Weill Cornell Medicine, New York, NY, United States

Contents

22.1 Introduction

Electromyography (EMG) is a diagnostic methodology employed in the assessment of neuromuscular health status. This procedure reviews nerve and muscular dysfunction and problems associated with neuromuscular signals. This technique stipulates the distinct features between myopathy and neurogenic muscular waste as well as weakness. EMG also detects fasciculate disorders. This investigation is usually used in cases

Practical Design and Applications of Medical Devices.
DOI: https://doi.org/10.1016/B978-0-443-14133-1.00007-0
299

of motor neuron disorders [1]. Neuromuscular disorders are common in disabled society, especially in indigent families whose incomes are low. The disabled or other amputees (as a result of accidents) are provided with different devices that have been established in the design of prosthetic hands [2]. These techniques often rely on the implementation of new actuator types, the development of kinematic structures that are more effective, as well as integration of effective compliance [3,4]. Presently available robotic prosthetic hands employ a single actuator type to perform the functions represented in the flexor digitorum superficialis and flexor digitorum profundus [5]. The powerful, lightweight, and silent shape memory actuators were chosen to offer appropriate system strength when necessary [6]. Because myoelectric prosthetics are very expensive, many amputees across the globe cannot afford prosthetics. Statistics have shown that prosthetics costs up to 15,000 dollars yet it cannot surpass or be fully compared with human hands [7]. The prosthetic hand allows the user only to grip. The functions of the EMG, as used in the design, aids in the change of position of these hands, although the user should be cautious in the handling or holding of things. Literature has shown that there are more advanced hands with superior functions, although they are very expensive [8]. According to Ziegler–Graham et al. (2008) the number of amputees across the globe are 1.7 million. In the United States about 50,000 to 100,000 [9] amputations occur on a daily basis [9]. Also, there are about 2,632,255 individuals with disabilities in the world right now, which is 6.4% of the total population [4], 20.8% of them have permanent changes that damage their body parts like hands and legs. So, making low-cost prosthetics meets an important need in the lives of a large proportion of low-income people who do not have the money to use promotional prostheses [10]. Computer-aided design innovations, as well as the integration of other technologies, have reduced the costs of prosthetic development and fabrication [11]. As a result, 3D printing machines have been employed in independent electric and mechanical placement for the fabrication of prosthetics, while hands are created on open source. In turn, a reduced microcontroller is constructed enabling EMG remote sensing and gadget activation. However, for the design of this gadget, an attempt was made to accomplish it with a five-stage phenomenological and empirical methodology, i.e., bibliographical research of the current state of the art, specialist advice, specification and selection of components, and, finally, testing and confirmation of the phase prototypes.

22.1.1 Prosthetic hand

The essence of the design of the prosthetic hand is to mimic the muscles, tendons, ligaments, joints, cartilage, and other connective tissues that make up the musculoskeletal system. This system gives the body its shape, keeps it stable, and lets it move. The skeletal system provides support, protection, and leverage to the skeletal system [12−14]. With the growth of biotechnology, innovation has spread to robotics and the creation of prosthetic hands. There are many types of prosthetic hands at present; some of them are fixed, and some are movable. The fixed ones are used to compensate for the loss of the limb in a cosmetic way only, without providing any movement. The movements can be categorized according to the type of control it has: mechanical or electronic [15]. The mechanical prosthetic hand can be made of metal, wood, or other materials and has joints that allow it to move by using mechanical movement systems (teeth, rollers, rotating axes, etc.) [16]. Nonetheless, this hand-driven EMG connected by electrodes aids movement through motors connected to the hand by a transmission mechanism that ensures a smooth and easy movement as possible. Electronic prosthetic hands can be controlled by several methods, including EMG signal detection, muscle action potential measurement, and various other systems [17].

22.2 Related articles

Many scholars have recently contributed their academic expertise in the electromechanics of hand-driven EMG signals. In the study conducted by Kyeong et al. [18] EMG signals were frequently used to help individuals with disabilities and render assistance to people to perform functions they would perform with their hands and fingers. Most research looked at the wrist and the whole arm, not just one finger at a time. The study focuses on machine learning methodology to make an EMG-based finger and hand expression classification model based on a fixed EMG signal. The results showed a higher ratio of gesticulations to channels than other related research using only time-domain characteristics. This suggests that the proposed technique can make the system easier to use and lessen the amount of work it has to do [19]. Also Claudio et al. [20] conducted their research on the multi-daily activities of mechanical hands controlled by EMG methodology. According to their study, since the 1960s, feed-forward regulation of energetic prosthetic arms has been done with the help of forearm EMG signals. Recent studies have demonstrated that it has the ability to regulate even a hand prosthesis with many joints, like

Touch Bionics' i-LIMB, together with a robotic hand with many fingers and varying degrees of freedom, employing models that have already been trained could help the patient learn how to use the hand more quickly [21]. Ana et al. [22] in a Robhand study using real-time EMG suggested that after a cerebro-vascular disorder, supported bilateral care has been demonstrated to improve the use of one's paralyzed limbs and speed up functional recovery, notably in the upper arms and legs. Other studies have looked at sensor bilateral recovery based on EMG-driven regulations as a way to enhance hand maneuverability. Nevertheless, low-cost implanted remedies for real-time are still needed. RobHand, a handmade, low-cost exoskeleton, helps with bilateral restoration. It sees how the healthy hand moves and then makes the same move on the robotic system on the paralyzed hand. A comprehensive review was done to contrast the outcomes of both mechanisms. The highest accuracy of the two methods for recognizing hand movements is 97%. These studies were limited to a singular methodology, but our unique design includes Arduino, motors, regulators, diodes, and an AT Mega 16 controller (Fig. 22.1).

22.3 Design methodology

22.3.1 Connection of electrical components to circuit boards

The design is made up of electrode patches, an Arduino, a servomotor, a 5-V external power source, and a regulator. The connection of all electronic and electrical equipment was performed in line with the design principle. A muscular detector was employed to detect the hand's bicep activity, while electrical cables and electrode patches complimented the installation. A torque-sensing servo, as well as Arduino R3 Control Board, were developed to create the motion. Arduino was used due to its flexibility, low price, and simple installation, as well as its comprehensive architecture with numerous power inputs, an Atmel AVR microprocessor, and a USB connection. Arduino UNO has proven its capacity to interpret a variety of biological inputs for biomedical fields, such as pulse rate monitoring [23] and EMG [23], among others. The MG995 servomotor has a great torque with durability and is meant to operate under heavy loads. The motor was chosen due to its application in biomedical projects [23] and its compatibility with the controller card [24]. The servomotor was fastened to the breadboard, which serves as a basis for the motor as well as the mechanical wiring. The servomotor is then placed on its own base.

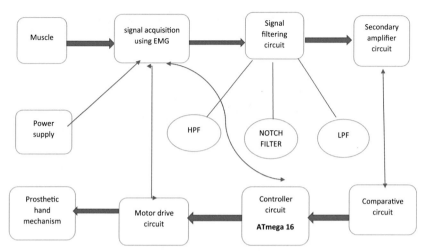

Figure 22.1 Shows the power circuit of EMG.

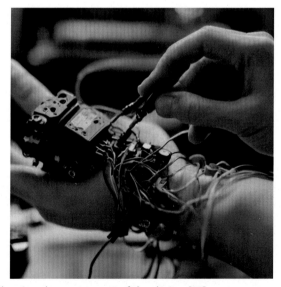

Figure 22.2 Showing the component of the design [25].

The design of the assemblage of these two pieces is seen in Fig. 22.2 The electrical connections were connected in accordance with the usual servo-motor connecting scheme and the company's instructions for the Arduino Uno board, using pulse-width modulation (PWM) output to drive the servomotor [16] and analog input from the Arduino for the signal's EMG sensors. To avoid limiting the servomotor current, a 5-V external source

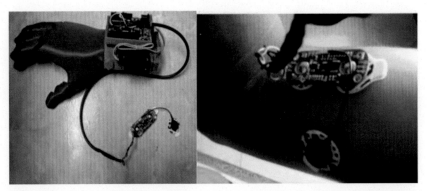

Figure 22.3 Shows the prototype of an Electromechanic hand driven EMG signal [26].

power source was used. Fig. 22.3 depicts the prototype's final construction with the sensor attached and implanted in the arm for functionality testing.

22.3.2 Amputation

Amputation is the process of removing one of the limbs of the human body, which may be for the lower or upper part of the limb. It can be a life-challenging procedure for the parts of the body that are removed [27]. A person who has undergone amputation can return to their normal life if they are given proper care. The term "prosthesis" refers to a prosthetic device that is used as a replacement for a missing part or an amputated part of the body [28]. After the amputation, the patient enters complex and unique psychological states due to the absence of a limb or part of it. During the amputation process, some nerves are cut, which causes the patient to feel that the limb is still there. It is necessary to go to a rehabilitation center within three to four weeks to start the process of replacing the amputated limb. This team will also select the most suitable limb for the patient at this stage, taking into consideration factors such as age, weight, and fitness [29]. Amputations are of two main types: lower limb and upper limb amputation. There are several factors that could cause amputation in human life, for example, dysvuascular, which is prevalent in developed countries. The vascular disease begins to appear in limbs far from the center of blood pumping. These limbs will suffer from a decrease in the amount of blood reaching them. 3% of the percentage of cases for this type of amputation for the upper limb and over 81% for the lower limb [30]. Other factors are trauma which accidents are the main cause of amputation. In third-world countries, the first cause of accidents that lead to amputation is war and it rates 6.68% for the upper limb and 31% for the lower limb. When this happens, the amputation should always be put off as long as possible because it's hard to tell how much of the affected tissue is still alive [31–33]. However, in cancer majority (75%) of its amputations occur above the knee [31] while congenital out of every 2000 births are cases of missing limbs congenitally. 68% are congenital and 32% are acquired amputations. Then, later in life, 62.5% are as a result of neoplastic issues, 18.8% were traumatically originated, 12.5% were a result of infectious, and 6.3% were vascularized [34].

22.3.3 The muscular system

The muscles are very important in the construction of hand-driven EMG signals. The contraction and relaxation as well as support to the structure of the human body give the body its general appearance as well as aid in locomotion [35]. These muscles are classified according to their shape. Flat muscles have parallel fibers, such as the external oblique abdominal muscle. The feather muscles whose bundles are arranged in the form of a feather may be single, bilateral, or multiple for instance the deltoid. The spindle muscle is spindle in shape with tapered ends like the brachialis. Circular muscles surround an opening in the body and narrow it when contracting [36]. Smooth muscle microscopically shows no cross stripes, [37] so it is a non-striated muscle, unlike skeletal muscle, which is striated. Smooth muscle derives its innervation from the autonomic nervous system and is therefore an involuntary muscle that frequently contracts [38]. Cardiac muscle contractions are not under voluntary control and are typical of striated muscles due to their appearance. Cardiac muscle fibers are chains of cardiac muscle cells with muscle junctions between them. There is an internal regulation of the heart rhythm by a pacemaker made up of specialized cardiac muscle fibers [39,40].

22.3.4 Methods and models of electric or electronic hand control

Electronic prosthetic hands can be controlled by several methods, including EMG signal detection, muscle action potential measurement, and various other systems. This project focuses on electrical muscle signals. This depicts the practical methods for designing the signal acquisition using the EMG circuit control signal, as well as the programming of the controller circuit for controlling the prosthetic hand.

The circuit has a voltage of about 5 v and it has about 3 diodes, viz. D4, D2, and D1, all have a parallax connection so as to protect the circuit from reversing power risk. This influx of voltage is controlled by a voltage regulator of 5 V. in order to get the desired voltage in control as well as the circuit. The supplier voltage is used, as illustrated in Fig. 22.2. The advantage of using L7805 is its ability to short circuit and overload protection as well as variability in voltage and 5.1 A output current. The MC34063 has the ability to convert DC to AC because of its several features, such as a wide input voltage, which can be varied range between 3 and 40 volts, and a frequency of 100 kHz.

22.3.5 Signal detection electrodes

Vital signs cannot indicate efficient functions in the human body without specialized electrodes that are suitable for vital signs. The electrodes have insulated wires which provide a cushion effect on the noise generated by the surroundings. Theoretically, there are different electrodes, but in view of this project, surface electrodes are used. They are used because they are readily available, give accurate results, and do not require clinical expertise or knowledge. The surface electrode also does not pose any danger to the recipient.

The signals from the muscles are captured by the electrodes. These impulses need to be amplified before being processed. It has about 30 to 300 microvolts. This initial amplification stage includes a differential amplification procedure, and the INA118P amplifier is used in this project. The signal muscle circuit has a common mode rejection ratio (CMRR) with high impedance as well as a short distance from the detection point. This is usually a universal amplifier that has a good track record for being accurate and using little power. It has been used in different applications, like the thermocouple. medical equipment as well as data acquisition. The design is unique in that it suits many applications. This amplifier has a wide bandwidth because of its stream backlight. The amplifier is always protected from the influx of high voltages. Internally, the structure is over voltage protected, as can be seen from the prototype. The common–mode gain to differential–mode gain (CMRR) ratio of the amplifiers used in detecting the vital signal, no matter how high its value, is not sufficient to get rid of the large common mode signal. This electro–circuit uses signal detection features to capture the mode signal from the human body, then amplifies it and the signal is reflected via the amplifier connected to the inverter

22.3.6 Control circuit

The control loops are very important. So as to move the prosthetic hand in the desired direction, the controller ensures the muscle signal obtained after processing in the previous stages is entered into the controller as a control signal, and then the controller, based on the programming commands, the motors in the prosthetic hand are activated to move to the desired movement. An example of a controller is the ATmega16. This is made up of several features, including 32 records. 8 of these 32 are used generally with an execution speed of about 16 multi-directional impact protection (MIPs) and a 16KB memory. The controller also contains 1 KB of data memory and an electrically erasable programmable read-only memory (EEPROM) with a capacity of 512 bytes. There are always four channels of PWM wave generation. Again, the controller has 3 timers, of which 2 are 8-bit and the other 16-bit (Fig. 22.4).

From the above diagram of the ATmega 16 port distribution, gates PA7-PA0 are used as input and output gates, respectively, with digital analog switches (Fig. 22.5).

The DC motors that are connected with the fingers to secure their movement need special circuits to convey the commands of the controller in an understandable way. L298 integrated circuit is used, each integrated circuit (IC) dedicated to driving two motors, so three of them is needed to drive the five motors we have on hand. The integrated circuit L298 is important because the current sensor within it, which enables us to calculate the resistive torque on the motor and thus determine the force of pressure of the fingers by connecting this sensor to the microcontroller. It has also by its ability to drive the

Figure 22.4 Shows the ATmega16 controller and port distribution.

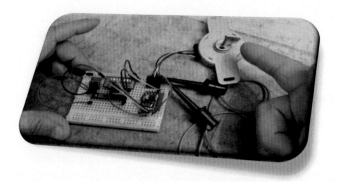

Figure 22.5 Motor drive circuit.

motor in two directions, ie reverse the direction of rotation of the motor and it can withstand a current of up to 2 A. The controller does all the job and this enables us to control the speed and strength of the movement of the fingers.

22.4 Conclusion

The most important part of prosthetic collaborative effort systems is the design of continuous human EMG support. EMG Signals infers human intentions. The EMG signal of a person's upper extremity could be used to predict a number of parameters, which would then be sent to the robot to help it do tasks that were planned. The study design enforces the muscles are captured by the electrodes and impulses were transmitted which are then amplified before being processed. This initial amplification stage includes a differential amplification procedure, and the INA118P amplifier is used in this project. The prototype is a replica of the muscles are being activated by the design and the electromechanics hand displays a mobile movement. this is a good boost especially low–income personnel's who are disabled or amputated. the study demonstrates excellent academic study and should not replace the standardize methodology of production.

References

[1] K.R. Mills, The basics of electromyography, Neurology in Practice 76 (2) (2005). Available from: https://doi.org/10.1136/JNNP.2005.069211.
[2] L.E. Osborn, M.M. Iskarous, N.V. Thakor, Sensing and control for prosthetic hands in clinical and research applications, Wearable Robotics: Systems and Applications (2019) 445–468. Available from: https://doi.org/10.1016/B978-0-12-814659-0.00022-9.

[3] H. Su, et al., Pneumatic soft robots: challenges and benefits, Actuators 11 (3) (2022) 92. Available from: https://doi.org/10.3390/ACT11030092.

[4] (PDF) Design of a robotic hand and simple EMG input controller with a biologically-inspired parallel actuation system for prosthetic applications. https://www.researchgate.net/publication/277064253_Design_of_a_robotic_hand_and_simple_EMG_input_controller_with_a_biologically-inspired_parallel_actuation_system_for_prosthetic_applications (accessed July 07, 2022).

[5] L. Dunai, M. Novak, C.G. Espert, Human hand anatomy-based prosthetic hand, Sensors (Basel) 21 (1) (2021) 1–15. Available from: https://doi.org/10.3390/S21010137.

[6] T. Senac, et al., A review of pneumatic actuators used for the design of medical simulators and medical tools a review of pneumatic actuators used for the design of medical simulators and medical tools, Multimodal Technologies and Interaction 3 (3) (2019) 47. Available from: https://doi.org/10.3390/mti3030047ï.

[7] This high schooler invented a low-cost, mind-controlled prosthetic arm, Innovation Smithsonian Magazine. https://www.smithsonianmag.com/innovation/this-high-schooler-invented-a-low-cost-mind-controlled-prosthetic-arm-180979984/ (accessed August 24, 2022).

[8] D.U. Ozsahin, et al., Designing a 3D printed artificial hand, Modern Practical Healthcare Issues in Biomedical Instrumentation (2021) 3–18. Available from: https://doi.org/10.1016/B978-0-323-85413-9.00009-8.

[9] K. Ziegler-Graham, E.J. MacKenzie, P.L. Ephraim, T.G. Travison, R. Brookmeyer, Estimating the prevalence of limb loss in the United States: 2005 to 2050, Archives of Physical Medicine and Rehabilitation 89 (3) (2008) 422–429. Available from: https://doi.org/10.1016/J.APMR.2007.11.005.

[10] Caracterización sobre Discapacidad a Nivel Nacional. https://docplayer.es/amp/75515164-Caracterizacion-sobre-discapacidad-a-nivel-nacional.html (accessed August 24, 2022).

[11] S.D. Rosca, M. Leba, A.F. Panaite, Modelling and simulation of 3D human arm prosthesis, Advances in Intelligent Systems and Computing 1160 (2020) 775–785. Available from: https://doi.org/10.1007/978-3-030-45691-7_73. AISC.

[12] N.V. Bhagavan, C.-E. Ha, Connective tissue, Essentials of Medical Biochemistry (2015) 119–136. Available from: https://doi.org/10.1016/B978-0-12-416687-5.00010-5.

[13] Anatomy, connective tissue, StatPearls—NCBI Bookshelf. https://www.ncbi.nlm.nih.gov/books/NBK538534/ (accessed July 07, 2022).

[14] R.A. Levine, Y. Oron, Tinnitus, Handbook of Clinical Neurology 129 (2015) 409–431. Available from: https://doi.org/10.1016/B978-0-444-62630-1.00023-8.

[15] A. Solgajová, T. Sollár, G. Vörösová, Original research article, Kontakt 2 (17) (2015) e67–e72. Available from: https://doi.org/10.1016/J.KONTAKT.2015.01.005.

[16] N. Tiep, N. Nguyen, Developing Low-cost Myoelectric Prosthetic Hand 47 pages + 2 appendices 18, 2018.

[17] S. Zimnowodzki, Concentric needle EMG, Encyclopedia of Movement Disorders (2010) 249–250. Available from: https://doi.org/10.1016/B978-0-12-374105-9.00102-7.

[18] S. Kyeong, et al., Surface electromyography characteristics for motion intention recognition and implementation issues in lower-limb exoskeletons, International Journal of Control, Automation and Systems (2022). Available from: https://doi.org/10.1007/s12555-020-0934-3.

[19] K.H. Lee, J.Y. Min, S. Byun, Electromyogram-based classification of hand and finger gestures using artificial neural networks, Sensors (Basel) 22 (1) (2022). Available from: https://doi.org/10.3390/S22010225.

[20] C. Claudio, F. Angelo Emanuele, S. Giulio, Multi-subject/daily-life activity EMG-based control of mechanical hands, Journal of NeuroEngineering and Rehabilitation 6 (2009)4110.1186/1743-0003-6-41.

[21] C. Castellini, A.E. Fiorilla, G. Sandini, Multi-subject/daily-life activity EMG-based control of mechanical hands, Journal of Neuroengineering and Rehabilitation 6 (1) (2009). Available from: https://doi.org/10.1186/1743-0003-6-41.

[22] C. Ana, RobHand: A hand exoskeleton with real-time EMG-driven embedded control. quantifying hand gesture recognition delays for bilateral rehabilitation, IEEE 9 (2021)10.1109/ACCESS.2021.3118281.

[23] H. Yang, C. Xiang, H. Han, L. Hao, Inverse kinematics modeling and motion control of PAM bionic elbow joint, in: 2015 IEEE International Conference on Robotics and Biomimetics, IEEE-ROBIO 2015, 2015, pp. 1347–1352. Available from: https://doi.org/10.1109/ROBIO.2015.7418958.

[24] Volume measuring system using arduino for automatic liquid filling machine. https://www.researchgate.net/publication/329683787_Volume_measuring_system_using_arduino_for_automatic_liquid_filling_machine (accessed August 24, 2022).

[25] Mike, Amazing 3D printed models on display at RSNA 2014, 3D Printing in Medicine (2014).

[26] C.L. Sandoval-Rodriguez, E.Y. Veslin-Díaz, B.E. Tarazona-Romero, J.G. Ascanio-Villabona, C.G. Cárdenas-Arias, C.A. Angulo-Julio, Electromechanical hand prototype for the simulation of the opening and closing movement, IOP Conference Series: Materials Science and Engineering 1154 (1) (2021) 012035. Available from: https://doi.org/10.1088/1757-899x/1154/1/012035.

[27] Amputation, Johns Hopkins Medicine. https://www.hopkinsmedicine.org/health/treatment-tests-and-therapies/amputation (accessed July 07, 2022).

[28] R. Bekrater-Bodmann, Perceptual correlates of successful body–prosthesis interaction in lower limb amputees: psychometric characterisation and development of the Prosthesis Embodiment Scale, Scientific Reports 10 (1) (2020) 1–13. Available from: https://doi.org/10.1038/s41598-020-70828-y.

[29] A.C. Roşca, C.C. Baciu, V. Burtăverde, A. Mateizer, Psychological consequences in patients with amputation of a limb. An interpretative-phenomenological analysis, Front Psychology 12 (2021) 537493. Available from: https://doi.org/10.3389/FPSYG.2021.537493.

[30] J. Kulkarni, S. Pande, J. Morris, Survival rates in dysvascular lower limb amputees, International Journal of Surgery (London, England) 4 (4) (2006) 217–221. Available from: https://doi.org/10.1016/J.IJSU.2006.06.027.

[31] M. Davies, L. Burdett, F. Bowling, N. Ahmad, J. Mcclennon, The epidemiology of major lower-limb amputation in England: a systematic review highlighting methodological differences of reported trials article points, The Diabetic Foot Journal 22 (2019) 4.

[32] G. Das Pooja, L. Sangeeta, Prevalence and aetiology of amputation in Kolkata, India: a retrospective analysis, Hong Kong Physiotherapy Journal 31 (1) (2013) 36–40. Available from: https://doi.org/10.1016/J.HKPJ.2012.12.002.

[33] P.L. Chalya, et al., Major limb amputations: a tertiary hospital experience in northwestern Tanzania, Journal of Orthopaedic Surgery and Research 7 (2012) 1. Available from: https://doi.org/10.1186/1749-799X-7-18.

[34] J.P. Fonseca, P. Figueiredo, P. Lemos Pereira, The role of rehabilitation in pediatric amputation — a 10-year retrospective study in a Portuguese population, International Physical Medicine & Rehabilitation Journal 7 (1) (2022) 21–24. Available from: https://doi.org/10.15406/ipmrj.2022.07.00298.

[35] K. Mukund, S. Subramaniam, Skeletal muscle: a review of molecular structure and function, in health and disease, Wiley Interdisciplinary Reviews: Systems Biology and Medicine 12 (1) (2020) e1462. Available from: https://doi.org/10.1002/WSBM.1462.

[36] Linea alba: attachments, relations and function, Kenhub. https://www.kenhub.com/en/library/anatomy/linea-alba (accessed July 07, 2022).

[37] C. Otero-Sabio, et al., Microscopic anatomical, immunohistochemical, and morphometric characterization of the terminal airways of the lung in cetaceans, Journal of Morphology 282 (2) (2021) 291–308. Available from: https://doi.org/10.1002/JMOR.21304.

[38] T. Brunet, A.H.L. Fischer, P.R.H. Steinmetz, A. Lauri, P. Bertucci, D. Arendt, The evolutionary origin of bilaterian smooth and striated myocytes, Elife 5 (2016). Available from: https://doi.org/10.7554/ELIFE.19607.

[39] H.L. Sweeney, D.W. Hammers, Muscle contraction, Cold Spring Harbor Perspectives in Biology 10 (2) (2018). Available from: https://doi.org/10.1101/CSHPERSPECT.A023200.

[40] D.D. Tang, The dynamic actin cytoskeleton in smooth muscle, Advances in Pharmacology 81 (2018) 1–38. Available from: https://doi.org/10.1016/BS.APHA.2017.06.001.

Index

Note: Page numbers followed by "*f*" and "*t*" refer to figures and tables, respectively.

P

Pacing pulse generator, 278
Pandas library, 120
Passive functional hand prostheses, 2–3
Patient well-being monitoring system,
    IoT-based
  challenges and drawbacks, 37–38
  components of, 27
  description of, 24
  implementation of, 33–35
  liquid crystal display
    connections with, 33
    connections without, 33
  proposed devices, 25–26
  system software performance, 36–37
  ThingSpeak, 31–32
  usefulness of, 25–26
  working principles, 27–28
    Arduino Uno, 28
    ESP8266, 30
    LCD, 30–31
    potentiometer, 31
    pulse rate sensor, 28–29
    temperature sensor, 29–30
PCR. *See* Polymerase chain reaction
    (PCR)
Peripheral vascular diseases, 100
Peristaltic pump, 222–223
Peritoneal dialysis, 221
Personal Disease Information Access
    System, 119–120
PET. *See* Positron emission tomography
    (PET)
Pharynx, 150
Photoplethysmography (PPG), 292–294
Physiotherapy, 87–89. *See also*
    Electrophysiotherapy
  categories of, 88–89
  electrostimulants, 89
PIC18F4520 microchip, 280–282
PIC18F4520 processor, 275
PIC16F87X microcontroller, 92–93
Piezoelectric accelerometer, 276–277
Piezo transducer, 156–157
Polymerase chain reaction (PCR)
  characteristics of, 76–77

deoxyribonucleic acid, 73–74
  device components, 75
  methodologies
    components, 80–82
    experimental design, 77–80
  overview of, 73–74
  real-time, 76–77
  stages in, 74–75
  types of, 75–76
Positron emission tomography (PET), 175
Potentiometer, 31
PPG. *See* Photoplethysmography (PPG)
Prostheses. *See also* 3D-design printed
    prosthetic legs
  3D printing of, 109
  importance of, 109
Prosthetic arm, voice-controlled
  artificially made body organs,
    development of, 101–102
  electromyogram used in, 6–7
  design and implementation, 102–103
  materials used in, prototype design,
    103–105
  overview of, 99–100
  peripheral vascular diseases, 100
Prototypes
  Covid-19 emergency medical kit, 67
  polymerase chain reaction, 82
Pulse generators, 278
Pulse oximeter mechanism, 293
  ambulatory blood pressure, 296
  design approach, 293–294
  operation methodology, 295–296
Pulse rate sensor, 28–29
Pulse sensor oximeter, 63–65
Pulse wave transit time (PWTT)
    approaches, 293–294, 296
Pulse-width modulation (PWM), 302–304
PWM. *See* Pulse-width modulation
    (PWM)
PWTT approaches. *See* Pulse wave transit
    time (PWTT) approaches

Q

Qualitative PCR, 75–76